ANALYTICAL CHEMISTRY

Edited by **Ira S. Krull**

Analytical Chemistry
http://dx.doi.org/10.5772/3086
Edited by Ira S. Krull

Contributors

Christophe B.Y. Cordella, S. A. Akinyemi, W. M. Gitari, A. Akinlua, L. F. Petrik, Njegomir Radić, Lea Kukoc-Modun, Miguel Valcárcel, Izydor Apostol, Ira Krull, Drew Kelner, Ion Neda, Paulina Vlazan, Raluca Oana Pop, Paula Sfarloaga, Ioan Grozescu, Adina-Elena Segneanu

Published by InTech

Janeza Trdine 9, 51000 Rijeka, Croatia

Publishing Process Manager Marina Jozipovic
Technical Editor InTech DTP team
Cover InTech Design team

Additional hard copies can be obtained from orders@intechopen.com

Analytical Chemistry, Edited by Ira S. Krull
p. cm.
ISBN 978-953-51-0837-5

For EU product safety concerns:
IN TECH d.o.o., Prolaz Marije Krucifikse Kozulić 3, 51000 Rijeka, Croatia,
info@intechopen.com or visit our website at intechopen.com.

Meet the editor

Ira S. Krull, obtained B.S. degree from City College of New York, NYC in 1962, and M.S, Ph.D from New York University, NYC in 1966 and 1968. His past appointments include: Senior Scientist at Boyce Thompson Institute; Adjunct Lecturer at NY Institute of Technology; Senior Scientist at ThermoElectron Corp; and Faculty Fellow and Associate Professor at Northeastern University. Currently, Prof. Krull is Professor Emeritus at Northeastern University. He has published more than 100 papers, and served as an editorial reviewer for many journals. Prof. Krull's research interests are in several areas of analytical biochemistry/biotechnology, especially methods for the analysis and detection of biopolymers and largely emphasize the use of separations methods including HPLC, HPCE, mass spectrometry, and capillary electrochromatography. He has developed, validated, and applied such separation methods for numerous proteins, antibodies, and peptides, of natural and/or recombinant derivation, especially those of commercial interest and medicinal applications.

Contents

Preface

The book, Analytical Chemistry, deals with some of the more important areas of the title's subject matter, which we hope will enable our readers to better appreciate just some of the numerous topics of importance in Analytical Chemistry today. The topics being covered herein deal with the following topics: 1) Principal Component Analysis: The Basic Building Block of Chemometrics; 2) Mineralogy and Geochemistry of Sub-bituminous Coal; 3) Kinetic Methods of Analysis with Potentiometric and Spectrophotometric Detectors; 4) Analytical Chemistry - Today and Tomorrow, Biochemical and Chemical Information, Basic Standards, Analytical Properties and Challenges; 5) Analytical Method Validation in Biotechnology Today, Method Validation, ICH/FDA Requirements, and Related Topics; and 6) Peptide and Amino Acid Separations and Identification - Chromatography, Spectrometry and Other Detection Methods. These topics are up-to-date, thoroughly referenced and literature cited to the present time, and truly review the major and most important topics and areas in each of these subjects. They should find widespread acceptance and studious involvement on the part of many readers interested in such generally important areas of Analytical Chemistry today. Analytical Chemistry has come to assume a tremendously important part of many, if not most, research areas in chemistry, biochemistry, biomedicine, medical science, proteomics, peptidomics, metabolomics, protein/peptide analysis, forensic science, and innumerable other areas of current, scientific research and development in modern times. As such, it behooves all those involved in these and other research and development areas to keep themselves up-to-date and current in the very latest developments in these and other areas of Analytical Chemistry.

Ira S. Krull
Chemistry and Chemical Biology,
Northeastern University,
Boston,
USA

Chapter 1

PCA: The Basic Building Block of Chemometrics

Christophe B.Y. Cordella

Additional information is available at the end of the chapter

http://dx.doi.org/10.5772/51429

1. Introduction

Modern analytical instruments generate large amounts of data. An infrared (IR) spectrum may include several thousands of data points (wave number). In a GC-MS (Gas Chromatography-Mass spectrometry) analysis, it is common to obtain in a single analysis 600,000 digital values whose size amounts to about 2.5 megabytes or more. There are different methods for dealing with this huge quantity of information. The simplest one is to ignore the bulk of the available data. For example, in the case of a spectroscopic analysis, the spectrum can be reduced to maxima of intensity of some characteristic bands. In the case analysed by GC-MS, the recording is, accordingly, for a special unit of mass and not the full range of units of mass. Until recently, it was indeed impossible to fully explore a large set of data, and many potentially useful pieces of information remained unrevealed. Nowadays, the systematic use of computers makes it possible to completely process huge data collections, with a minimum loss of information. By the intensive use of chemometric tools, it becomes possible to gain a deeper insight and a more complete interpretation of this data. The main objectives of multivariate methods in analytical chemistry include data reduction, grouping and the classification of observations and the modelling of relationships that may exist between variables. The predictive aspect is also an important component of some methods of multivariate analysis. It is actually important to predict whether a new observation belongs to any pre-defined qualitative groups or else to estimate some quantitative feature such as chemical concentration. This chapter presents an essential multivariate method, namely *principal component analysis* (PCA). In order to better understand the fundamentals, we first return to the historical origins of this technique. Then, we will show - via pedagogical examples - the importance of PCA in comparison to traditional univariate data processing methods. With PCA, we move from the one-dimensional vision of a problem to its multidimensional version. Multiway extensions of PCA, PARAFAC and Tucker3 models are exposed in a second part of this chapter with brief historical and bibliographical elements. A PARAFAC example on real data is presented in order to illustrate the interest in this powerful technique for handling high dimensional data.

2. General considerations and historical introduction

The origin of PCA is confounded with that of linear regression. In 1870, Sir Lord Francis Galton worked on the measurement of the physical features of human populations. He assumed that many physical traits are transmitted by heredity. From theoretical assumptions, he supposed that the height of children with exceptionally tall parents will, eventually, tend to have a height close to the mean of the entire population. This assumption greatly disturbed Lord Galton, who interpreted it as a "move to mediocrity," in other words as a kind of *regression* of the human race. This led him in 1889 to formulate his law of *universal regression* which gave birth to the statistical tool of *linear regression*. Nowadays, the word "regression" is still in use in statistical science (but obviously without any connotation of the regression of the human race). Around 30 years later, Karl Pearson,[1] who was one of Galton's disciples, exploited the statistical work of his mentor and built up the mathematical framework of *linear regression*. In doing so, he laid down the basis of the correlation calculation, which plays an important role in PCA. Correlation and linear regression were exploited later on in the fields of psychometrics, biometrics and, much later on, in chemometrics.

Thirty-two years later, Harold Hotelling[2] made use of correlation in the same spirit as Pearson and imagined a graphical presentation of the results that was easier to interpret than tables of numbers. At this time, Hotelling was concerned with the *economic games* of companies. He worked on the concept of economic competition and introduced a notion of spatial competition in *duopoly*. This refers to an economic situation where many "players" offer similar products or services in possibly overlapping trading areas. Their potential customers are thus in a situation of choice between different available products. This can lead to illegal agreements between companies. Harold Hotelling was already looking at solving this problem. In 1933, he wrote in the *Journal of Educational Psychology* a fundamental article entitled "Analysis of a Complex of Statistical Variables with Principal Components," which finally introduced the use of special variables called *principal components*. These new variables allow for the easier viewing of the intrinsic characteristics of observations.

PCA (also known as Karhunen-Love or Hotelling transform) is a member of those descriptive methods called *multidimensional factorial methods*. It has seen many improvements, including those developed in France by the group headed by Jean-Paul Benzécri[3] in the 1960s. This group exploited in particular geometry and graphs. Since PCA is

---

[1] Famous for his contributions to the foundation of modern statistics ($\chi 2$ test, linear regression, principal components analysis), Karl Pearson (1857-1936), English mathematician, taught mathematics at London University from 1884 to 1933 and was the editor of *The Annals of Eugenics* (1925 to 1936). As a good disciple of Francis Galton (1822-1911), Pearson continues the statistical work of his mentor which leads him to lay the foundation for calculating correlations (on the basis of principal component analysis) and will mark the beginning of a new science of biometrics.

[2] Harold Hotelling: American economist and statistician (1895-1973). Professor of Economics at Columbia University, he is responsible for significant contributions in statistics in the first half of the 20th century, such as the calculation of variation, the production functions based on profit maximization, the using the t distribution for the validation of assumptions that lead to the calculation of the confidence interval.

[3] French statistician. Alumnus of the Ecole Normale Superieure, professor at the Institute of Statistics, University of Paris. Founder of the French school of data analysis in the years 1960-1990, Jean-Paul Benzécri developed statistical

a descriptive method, it is not based on a probabilistic model of data but simply aims to provide geometric representation.

3. PCA: Description, use and interpretation of the outputs

Among the multivariate analysis techniques, PCA is the most frequently used because it is a starting point in the process of data mining [*1, 2*]. It aims at minimizing the dimensionality of the data. Indeed, it is common to deal with a lot of data in which a set of n objects is described by a number p of variables. The data is gathered in a matrix X, with n rows and p columns, with an element x_{ij} referring to an element of X at the ith line and the jth column. Usually, a line of X corresponds with an "observation", which can be a set of physicochemical measurements or a spectrum or, more generally, an analytical curve obtained from an analysis of a real sample performed with an instrument producing analytical curves as output data. A column of X is usually called a "variable". With regard to the type of analysis that concerns us, we are typically faced with multidimensional data n x p, where n and p are of the order of several hundreds or even thousands. In such situations, it is difficult to identify in this set any relevant information without the help of a mathematical technique such as PCA. This technique is commonly used in all areas where data analysis is necessary; particularly in the food research laboratories and industries, where it is often used in conjunction with other multivariate techniques such as discriminant analysis (Table 1 indicates a few published works in the area of food from among a huge number of publications involving PCA).

3.1. Some theoretical aspects

The key idea of PCA is to represent the original data matrix X by a product of two matrices (smaller) T and P (respectively the scores matrix and the loadings matrix), such that:

$$ {}_{n}X^{p} = {}_{n}T^{q} \cdot {}_{p}\left[P^{t}\right]^{q} + {}_{n}E^{p} \tag{1}$$

Or the non-matrix version:

$$x_{ij} = \sum_{k=1}^{K} t_{ik} p_{kj} + e_{ij}$$

With the condition $p_i^t p_j = 0$ *and* $t_i^t t_j = 0$ *for* $i \neq j$

This orthogonality is a means to ensure the non-redundancy (at least at a minimum) of information "carried" by each estimated principal component.

Equation 1 can be expressed in a graphical form, as follows:

tools, such as correspondence analysis that can handle large amounts of data to visualize and prioritize the information.

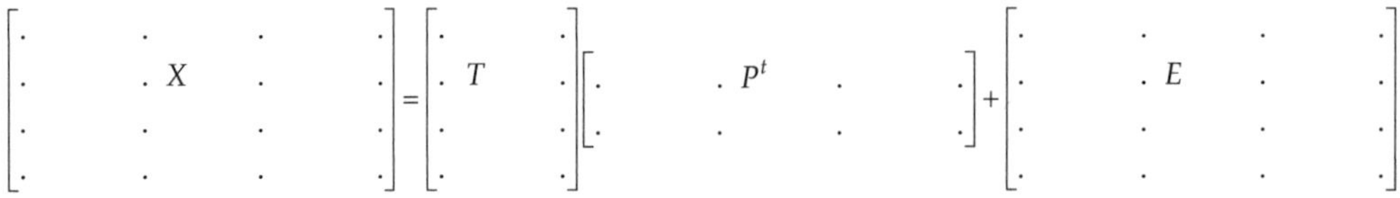

Scheme 1. A matricized representation of the PCA decomposition principle.

Schema 2 translates this representation in a vectorized version which shows how the X matrix is decomposed in a sum of column-vectors (components) and line-vectors (eigenvectors). In a case of spectroscopic data or chromatographic data, these components and eigenvectors take a chemical signification which is, respectively, the proportion of the constituent *i* for the ith component and the "pure spectrum" or "pure chromatogram" for the ith eigenvector.

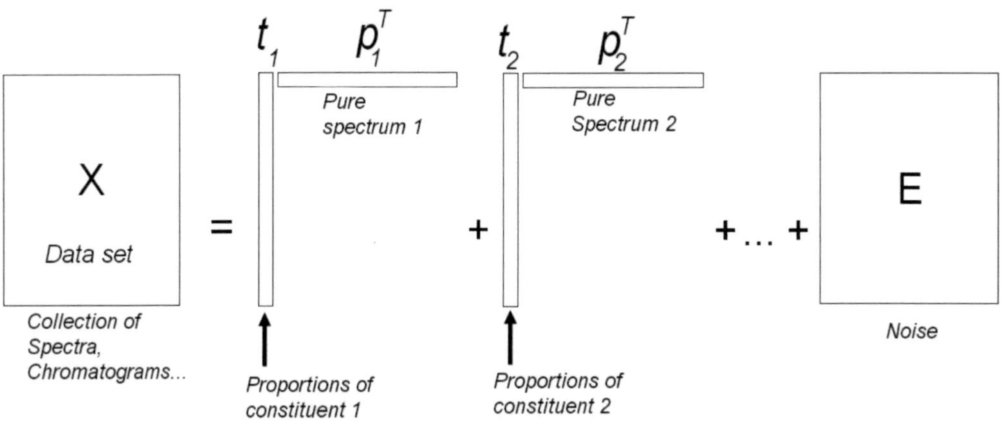

Scheme 2. Schematic representation of the PCA decomposition as a sum of "components" and "eigenvectors" with their chemical significance.

The mathematical question behind this re-expression of *X* is: is there another basis which is a linear combination of the original basis that re-expresses the original data? The term "basis" means here a mathematical basis of unit vectors that support all other vectors of data. Regarding the linearity - which is one of the basic assumptions of the PCA - the general response to this question can be written in the case where *X* is perfectly re-expressed by the matrix product $T.P^T$, as follows:

$$PX = T \tag{2}$$

Equation 2 represents a *change of basis* and can be interpreted in several ways, such as the transformation of *X* into *T* by the application of *P* or by geometrically saying that *P* - which is the rotation and a stretch - transforms *X* into *T* or the rows of *P*, where $\{p_1,\ldots,p_m\}$ are a set of new basis vectors for expressing the columns of *X*.

Therefore, the solution offered by PCA consists of finding the matrix *P*, of which at least three ways are possible:

1. by calculating the eigenvectors of the square, symmetric covariance matrix X^TX (eigenvector analysis implying a diagonalization of the X^TX);
2. by calculating the eigenvectors of X by the direct decomposition of X using an iterative procedure (NIPALS);
3. by the singular value decomposition of X - this method is a more general algebraic solution of PCA.

The dual nature of expressions $T = PX$ and $X = TP^T$ leads to a comparable result when PCA is applied on X or on its transposed X^T. The score vectors of one are the eigenvectors of the other. This property is very important and is utilized when we compute the principal components of a matrix using the covariance matrix method (see the pedagogical example below).

3.2. Geometrical point of view

3.2.1. *Change of basis vectors and reduction of dimensionality*

Consider a p-dimensional space where each dimension is associated with a variable. In this space, each observation is characterized by its coordinates corresponding to the value of variables that describe it. Since the raw data is generally too complex to lead to an interpretable representation in the space of the initial variables, it is necessary to "compress" or "reduce" the p-dimensional space into a space smaller than p, while maintaining the maximum information. The amount of information is statistically represented by the variances. PCA builds new variables by the linear combination of original variables. Geometrically, this change of variables results in a change of axes, called *principal components,* chosen to be orthogonal[4]. Each newly created axis defines a direction that describes a part of the global information.

The first component (i.e. the first axis) is calculated in order to represent the main pieces of information, and then comes the second component which represents a smaller amount of information, and so on. In other words, the p original variables are replaced by a set of new variables, the *components,* which are linear combinations of these original variables. The variances of components are sorted in decreasing order. By construction of PCA, the whole set of components keeps all of the original variance. The dimensions of the space are not then reduced but the change of axis allows a better representation of the data. Moreover, by retaining the q first principal components (with $q<p$), one is assured to retain the maximum of the variance contained in the original data for a q-dimensional space. This reduction from p to q dimensions is the result of the projection of points in a p-dimensional space into a subspace of dimension q. A highlight of the technique is the ability to represent simultaneously or separately the samples and variables in the space of initial components.

[4] The orthogonality ensures the non-correlation of these axes and, therefore, the information carried by an axis is not even partially redundant with that carried by another.

Food(s)	Analysed compounds	Analytical technique(s)	Chemometrics	Aim of study	Year [*Ref.*]
Cheeses	Water-soluble compounds	HPLC	PCA, LDA	Classification	1990 [3]
Edible oils	All chemicals between 4800 et 800 cm^{-1}	FTIR (Mid-IR)	PCA, LDA	Authentication	1994 [4]
Fruit puree	All chemicals between 4000 et 800 cm^{-1}	FTIR (Mid-IR)	PCA, LDA	Authentication	1995 [5]
Orange juice	All chemicals between 9000 et 4000 cm^{-1}	FTIR (NIR)	PCA, LDA	Authentication	1995 [6]
Green coffee	All chemicals between 4800 et 350 cm^{-1}	FTIR (Mid-IR)	PCA, LDA	Origin	1996 [7]
Virgin olive oil	All chemicals between 3250 et 100 cm^{-1}	FT-Raman (Far- et Mid-Raman)	PCR, LDA, HCA	Authentication	1996 [8]
General[5]	General	FTIR (Mid-IR)	PCA, LDA + others	Classification & Authentication	1998 [9]
Almonds	Fatty acids	GC	PCA	Origin	1996 [10]
			PCA, LDA		1998 [11]
Garlic products	Volatile sulphur compounds	GC-MS	PCA	Classification	1998 [11]
Cider apples fruits	Physicochemical parameters	Physicochemical techniques + HPLC for sugars	LDA	Classification	1998 [12, 13]
Apple juice	Aromas	Capillary GC-MS + chiral GC-MS	PCA	Authentication	1999 [14]
Meet	All chemicals between 25000 et 4000 cm^{-1}	IR (NIR+Visible)	PCA + others	Authentication	2000 [15]
Coffees	Physicochemical parameters + chlorogenic acid	Physicochemical techniques + HPLC for chlorogenic acid determination	PCA	Classification according to botanical origin and other criteria	2001 [16]
-	Data from various synthetic substances	MS + IR	PCA, HCA	Comparison of classification methods	2001 [17]
Honeys	Sugars	GC-MS	PCA, LDA	Classification according to floral origin	2001 [18]
Red wine	Phenolic compounds	HPLC	PCA (+PLS)	Relationship between phenolic compounds and antioxidant power	2001 [19]
Honeys	All chemicals between 9000-4000 cm-1	IR (NIR)	PCA, LDA	Classification	2002 [19]

Table 1. Examples of analytical work on various food products, involving the PCA (and/or LDA) as tools for data processing (from 1990 to 2002). Not exhaustive.

[5] Comprehensive overview on the use of chemometric tools for processing data derived from infrared spectroscopy applied to food analysis.

3.2.2. Correlation circle for discontinuous variables

In the case of physicochemical or sensorial data, more generally in cases where variables are not continuous like in spectroscopy or chromatography, a powerful tool is useful for interpreting the meaning of the axes: the correlation circle. On this graph, each variable is associated with a point whose coordinate on an axis factor is a measure of the correlation between the variable and the factor. In the space of dimension p, the maximum distance of the variables at the origin is equal to 1. So, by projection on a factorial plan, the variables are part of a circle of radius 1 (the correlation circle) and the closer they are near the edge of the circle, the more they are represented by the plane of factors. Therefore, the variables correlate well with the two factors constituting the plan.

The angle between two variables, measured by its cosine, is equal to the linear correlation coefficient between two variables: cos (angle) = r (V1, V2)

- if the points are very close (angle close to 0): cos (angle) = r (V1, V2) = 1 then V1 and V2 are very highly positively correlated
- if a is equal to 90 °, cos (angle) = r (V1, V2) = 0 then no linear correlation between X1 and X2
- if the points are opposite, a is 180 °, cos (angle) = r (V1, V2) = -1: V1 and V2 are very strongly negatively correlated.

An illustration of this is given by figure 1, which presents a correlation circle obtained from a principal component analysis on physicochemical data measured on palm oil samples. In this example, different chemical parameters (such as *Lauric acid* content, concentration of *saponifiable* compounds, iodine index, *oleic acid* content, etc.) have been measured. One can note for example on the PC1 axis, *"iodine index"* and *"Palmitic"* are close together and so have a high correlation; in the same way, *"iodine index"* and *"Oleic"* are positively correlated because they are close together and close to the circle. On the other hand, a similar interpretation can be made with *"Lauric"* and *"Miristic"* variables, indicating a high correlation between these two variables, which are together negatively correlated on PC1. On PC2, *"Capric"* and *"Caprilic"* variables are highly correlated. Obviously, the correlation circle should be interpreted jointly with another graph (named the score-plot) resulting from the calculation of samples coordinates in the new principal components space that we discuss below in this chapter.

3.2.3. Scores and loadings

We speak of **scores** to denote the **coordinates** of the observations on the PC components and the corresponding graphs (objects projected in successive planes defined by two principal components) are called score-plots. **Loadings** denotes the **contributions** of original variables to the various components, and corresponding graphs called **loadings-plot** can be seen as the projection of unit vectors representing the variables in the successive planes of the main components. As scores are a representation of observations in the space formed by the new axes (principal components), symmetrically, loadings represent the variables in the space of principal components.

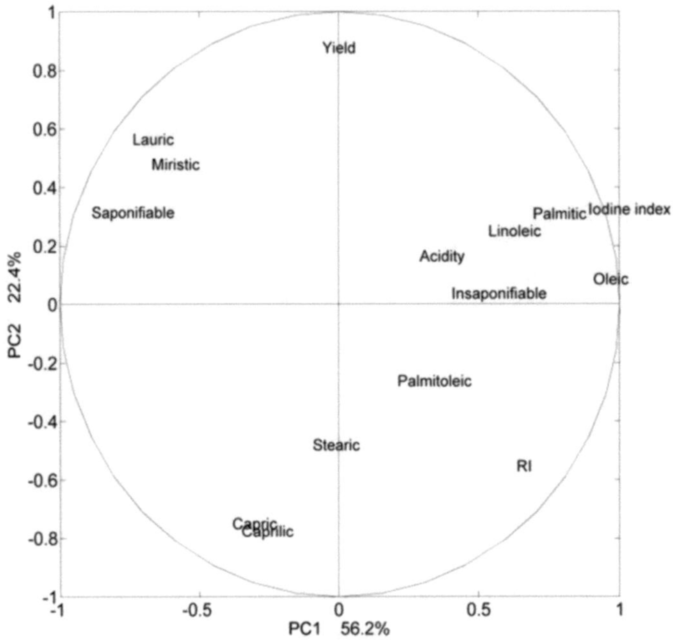

Figure 1. Example of score-plot and correlation circle obtained with PCA.

Observations close to each other in the space of principal components necessarily have similar characteristics. This proximity in the initial space leads to a close neighbouring in the score-plots. Similarly, the variables whose unit vectors are close to each other are said to be positively correlated, meaning that their influence on the positioning of objects is similar (again, these proximities are reflected in the projections of variables on loadings-plot). However, variables far away from each other will be defined as being negatively correlated.

When we speak of *loadings* it is necessary to distinguish two different cases depending on the nature of the data. When the data contains discontinuous variables, as in the case of physicochemical data, the loadings are represented as a factorial plan, i.e. PC1 vs. PC2, showing each variable in the PCs space. However, when the data is continuous (in case of spectroscopic or chromatographic data) loadings are not represented in the same way. In this case, uses usually represent the values of the loadings of each principal component in a graph with the values of the loadings of component PCi on the Y-axis and the scale corresponding to the experimental unit on the X-axis Thus, the loadings are like a spectrum or a chromatogram (See § B. Research example: Application of PCA on 1H-NMR spectra to study the thermal stability of edible oils). Figure 2 and Figure 3 provide an example of scores and loadings plots extracted from a physicochemical and sensorial characterization study of Italian beef [20] through the application of principal component analysis. The goal of this work was to discriminate between the ethnic groups of animals (hypertrophied Piemontese, HP; normal Piemontese, NP; Friesian, F; crossbred hypertrophied PiemontesexFriesian, HPxF; Belgian Blue and White, BBW). These graphs are useful for determining the likely reasons of groups' formation of objects that are visualized, i.e. the weight (or importance) of certain variables in the positioning of objects on the plane formed by two main components. Indeed, the objects placed, for example, right on the scores' plot will have important values for the variables placed right on the loadings plot, while the variables near the origin of the axes that will make a small contribution to the discrimination of objects. As presented by the authors of this work [20], one can see on the loadings plot that PC1 is characterized mainly by eating quality, one chemical parameter and two physical parameters (ease of sinking, Te; overall acceptability, Oa; initial juiciness, Ji; sustained juiciness, Js; friability, Tf; residue, Tr). These variables are located far from the origin of the first PC, to the right in the loadings plot, and close together, which means, therefore, that they are positively correlated. On the other hand, PC2 is mainly characterized by two chemical (hydroproline, Hy; and ether extract, E) and two physical (hue, H; and lightness, L) parameters. These variables located on the left of the loadings plot are positively correlated. The interpretation of the scores' plot indicates an arrangement of the samples into two groups: the first one includes meats of hypertrophied animals (HP) while the second one includes the meats of the normal Piemontese (NP) and the Friesian (F). Without repeating all of the interpretation reported by the authors, the combined reading of scores and loadings shows, for example, that the meat samples HP and BBW have, in general, a higher protein content as well as good eating qualities and lightness. At the opposite end, the meat samples F and NP are characterized more by their hydroxyproline content, their ether extract or else their Warner-Bratzler shear value. This interpretation may be made with the rest of the parameters studied and contributes to a better understanding of the influence of these parameters on the features of the product studied. This is a qualitative approach to comparing samples on the basis of a set of experimental measurements.

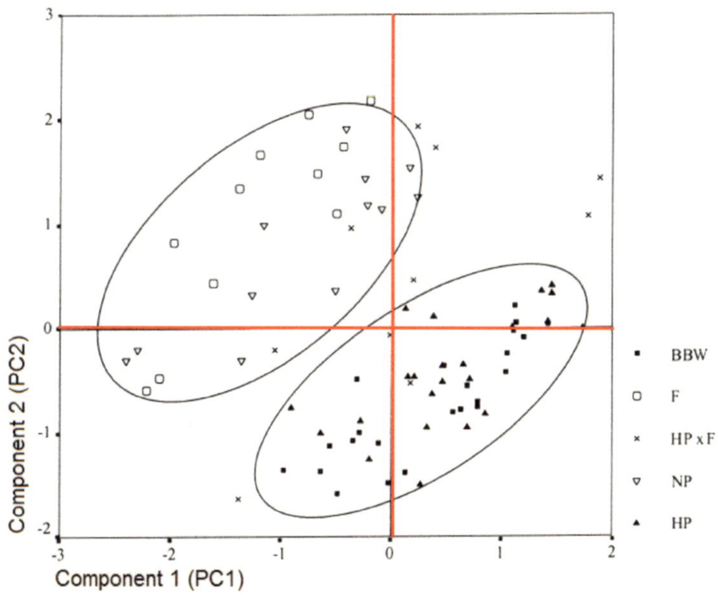

Figure 2. Score-plot obtained by PCA applied on meat samples. Extracted from [20].

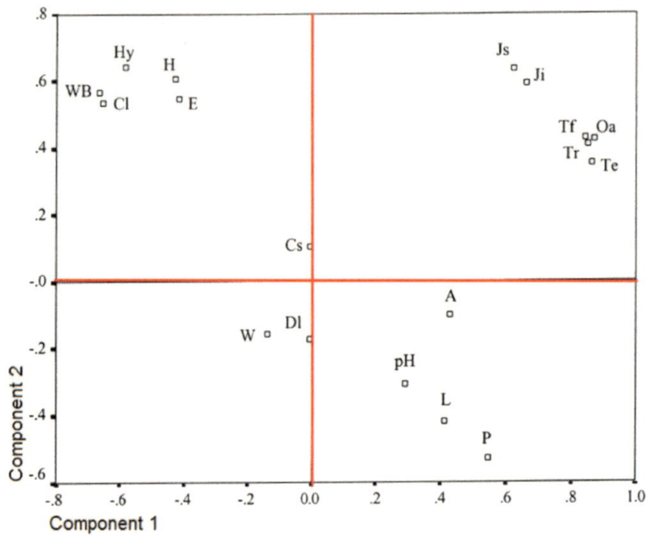

Plot of the first two PC loading vectors. Water (W); protein (P); ether extract (E); hydroxyproline (Hy); collagen solubility (Cs); lightness (L); hue (H); drip losses (Dl); cooking losses (Cl); Warner±Bratzler shear (WB); appearance (A); ease of sinking (Te); friability (Tf); residue (Tr); initial juiciness (Ji); sustained juiciness (Js); overall acceptability (Oa).

Figure 3. Loadings-plot obtained by PCA applied on meat samples. Extracted from [*20*].

3.3. Algorithms

There are different ways to achieve PCA, depending on whether one uses an iterative algorithm such as the NIPALS algorithm (Non-linear Iterative Partial Least Squares) or else a matrix factorization algorithm like SVD (Singular Value Decomposition). There are many variants of the SVD algorithm; the most well-known is probably the Golub-Reinsch algorithm [GR-SVD] [*21, 22*]. Both types of algorithms are well-suited for computing the eigenvalues and eigenvectors of a matrix *M* and determining, ultimately, the new axes of representation.

3.3.1. NIPALS

Let us start with the description of NIPALS. The most common version is given below.

M	Preferably the mean centred data matrix
$E(0) = M$	E-matrix equals to mean centred M at the beginning
t	Initialization step: this vector is set to a column in M, Scores for PC_i
p	Loadings for PC_i
$threshold = 10^{-4}$	Convergence criterion, a constant in the procedure

Beginning of the main loop (i=1 to Nb of PCs):

1. Project M onto t to calculate the corresponding loading p

$$p = \frac{E_{(i-1)}^T t}{t^T * t}$$

2. Normalise loading vector p to length 1

$$p = p * \sqrt{p^T * p}$$

3. Project M onto p to calculate the corresponding score vector t

$$t = \frac{E_{(i-1)}^T p}{p^T * p}$$

4. Check for convergence.

IF $\tau_{new} - \tau_{old} > threshold * \tau_{new}$ THEN return to step 1

With $\tau_{new} = (t^T t)$ and $\tau_{old} = (t^T t)$ from (i -1)th iteration.

5. Deflating process: remove the estimated PC component from $E_{(i-1)}$:

$$E_{(i)} = E_{(i-1)} - tp^T$$

End of the main loop

3.3.2. SVD

SVD is based on a theorem from linear algebra which says that a rectangular matrix M can be split down into the product of three new matrices:

- An orthogonal matrix U;
- A diagonal matrix S;
- The transpose of an orthogonal matrix V.

Usually the theorem is written as follows:

$$M_{mn} = U_{mm} S_{mn} V_{nn}^T$$

where $U^T U = I$; $V^T V = I$; the columns of U are orthonormal eigenvectors of MM^T, the columns of V are orthonormal eigenvectors of $M^T M$, and S is a diagonal matrix containing the square roots of eigenvalues from U or V in decreasing order, called singular values. The singular values are always real numbers. If the matrix M is a real matrix, then U and V are also real. The SVD represents an expression of the original data in a coordinate system where the covariance matrix is diagonal. Calculating the SVD consists of finding the eigenvalues and eigenvectors of MM^T and M^TM. The eigenvectors of M^TM will produce the columns of V and the eigenvectors of MM^T will give the columns of U.

Presented below is the pseudo-code of the SVD algorithm:

Compute M^T, M^TM (or MM^T)*

6. Compute eigenvalues of M^TM and sort them in descending order along its diagonal by resolving:

$$\left|M^T M - \lambda \mathrm{I}\right| = 0$$

- Characteristic equation: its resolution gives eigenvalues of M^TM.

1. Square root the eigenvalues of M^TM to obtain the singular values of M,
2. Build a diagonal matrix S by placing singular values in descending order along its diagonal and compute S^{-1},
3. Re-use eignenvalues from step 2 in descending order and compute the eigenvectors of M^TM. Place these eigenvectors along the columns of V and compute its transpose V^T.

Compute U as $U = MVS^{-1}$ (**) and compute the true scores T as $T = US$.

3.3.3. NIPALS versus SVD

The relationship between the matrices obtained by NIPALS and SVD is given by the following:

$$TP = USV$$

With the orthonormal US product corresponding to the scores matrix T, and V to the loadings matrix P. Note that with S being a diagonal matrix, its dimensions are the same as those of M.

4. Practical examples

4.1. Pedagogical example: how to make PCA step-by-step (See Matlab code in appendix)

The data in Table 2 consists of the fluorescence intensities at four different wavelengths for 10 hypothetical samples 1-10.

The data processing presented here was performed with Matlab v2007b. Like all data processing software, Matlab has a number of statistical tools to perform PCA in one mouse click or a one-step command-process, but we choose to give details of the calculation and bring up the steps leading to the representation of the factorial coordinates (scores) and factor contributions (loadings). Then, we interpret the results.

---

* Demonstrations can be found elsewhere showing that eigenvalues of $\boldsymbol{M^TM}$ and $\boldsymbol{MM^T}$ are the same. The reason is that $\boldsymbol{M^TM}$ and $\boldsymbol{MM^T}$ respond to the same characteristic equation.

** We know that $\boldsymbol{M = USV^T}$. Therefore, to find $\boldsymbol{U}$ knowing $\boldsymbol{S}$, just post-multiply by $\boldsymbol{S^{-1}}$ to obtain $\boldsymbol{AVS^{-1} = USS^{-1}}$. Hence $\boldsymbol{U = AVS^{-1}}$, because $\boldsymbol{SS^{-1} = I = 1}$.

Sample #	Wavelengths (nm)			
	300	**350**	**400**	**450**
1	15,4	11,4	6,1	3,6
2	16,0	12,1	6,1	3,4
3	14,0	13,1	7,3	4,2
4	16,7	10,4	5,9	3,1
5	17,1	12,1	7,1	2,9
6	81,9	71,3	42,8	32,9
7	94,9	85,0	50,0	39,0
8	69,9	75,7	46,3	50,4
9	70,4	70,5	42,6	42,1
10	61,8	81,9	50,2	66,2
Mean	15.75	61.25	68.92	29.25

Table 2. Fluorescence intensities for four wavelengths measured on 12 samples.

Step 1. Centring the data matrix by the average

This step is to subtract the intensity values of each column, the average of the said column. In other words, for each wavelength is the mean of all samples for this wavelength and subtract this value from the fluorescence intensity for each sample for the same wavelength.

Step 2. Calculation of the variance-covariance matrix

The variance-covariance matrix (Table 3) is calculated according to the $X^{T}X$ product, namely the Gram matrix. The diagonal of this matrix consists of the variances; the trace of the matrix (sum of diagonal elements) corresponds to the total variance (3294.5) of the original matrix of the data. This matrix shows, for example, that the covariance for the fluorescence intensities at 420 and 520 nm is equal to 665.9.

λ, nm	**420**	**474**	**520**	**570**
420	**1072,2**	1092,9	665,9	654,5
474	1092,9	**1194,2**	731,5	786,6
520	665,9	731,5	**448,4**	485,5
570	654,5	786,6	485,5	**579,8**

Table 3. Variance-covariance matrix

As mentioned above, the matrix also contains the variances of the fluorescence intensities at each wavelength on the diagonal. For example, for the fluorescence intensities at 474 nm, the variance is 1194.2.

The technique for calculating the eigenvectors of the variance-covariance matrix, and so the principal components, is called eigenvalue analysis (*eigenanalysis*).

Step 3. Calculate the eigenvalues and eigenvectors

Table 4 shows the eigenvalues and eigenvectors obtained by the diagonalization of the variance-covariance matrix (a process that will not be presented here, but some details on this mathematical process will be found elsewhere [23]).

Note that the sum of the eigenvalues is equal to the sum of the variances in the variance-covariance matrix, which is not surprising since the eigenvalues are calculated from the variance-covariance matrix.

	Eigenvalues		
3163,3645	0,0000	0,0000	0,0000
0,0000	130,5223	0,0000	0,0000
0,0000	0,0000	0,5404	0,0000
0,0000	0,0000	0,0000	0,1007
	Eigenvectors		
0,5662	0,6661	0,4719	0,1143
0,6141	-0,0795	-0,7072	0,3411
0,3760	-0,0875	-0,1056	-0,9164
0,4011	-0,7365	0,5157	0,1755

Table 4. Eigenvalues calculated using diagonalization processing of the variance-covariance matrix.

The next step of the calculation is to determine what percentage of the variance is explained by each major component. This is done by using the fact that the sum of the eigenvalues corresponding to 100% of the explained variance, as follows:

$$\%Var_{j^{th}\ PC} = \frac{\lambda_j}{\sum_{i=1}^{l} \lambda_i} \times 100\%$$

Where λ_j is the jth eigenvalue. We thus obtain for the first component PC1 = 3163.4 * 1 / 3294.5 * 100 = 96.01%. PC2 is associated with 3.96% and so on, as shown in Table 5 below.

PC1	PC2	PC3	PC4
96.02	3.96	0.02	0.00

Table 5. Percentage of the explained variance for each principal component.

Step 4. Calculating factorial coordinates - Scores

The eigenvectors calculated above are the principal components and the values given in Table 6 are the coefficients of each principal component. Thus, component #1 is written: PC1 = 0.566*X1 + 0.614*X2 + 0.376*X3 + 0.401*X4 where X1, X2, X3 and X4 are the fluorescence intensities, 420, 474, 520 and 570 nm respectively. The factorial coordinates of each sample in the new space formed by the principal components can now be calculated directly from the equations of the PCs. For example, sample No. 1 has a coordinate on PC1 equal to:

0.566*15.4 + 0.614*11.4 + 0.376*6.1 + 0.401*3.6 = 19.453. Table 6 presents the factorial coordinates of all the samples of the initial matrix.

	Centred data			
	PC1	PC2	PC3	PC4
1	19,455	6,128	0,377	0,678
2	20,121	6,671	0,072	0,983
3	20,420	4,528	-1,256	0,156
4	19,357	7,534	1,494	0,575
5	20,928	7,635	0,259	0,051
6	119,413	20,924	0,667	0,179
7	140,394	23,343	-0,522	0,843
8	123,676	-0,601	0,537	0,228
9	116,052	6,511	0,590	0,397
10	130,754	-18,524	0,056	0,632

Table 6. Factorial coordinates = *scores* of the initial sample matrix in the principal components space.

Therefore, we can now represent, for example, all samples in the space of two first components, as shown in Figure 4 below.

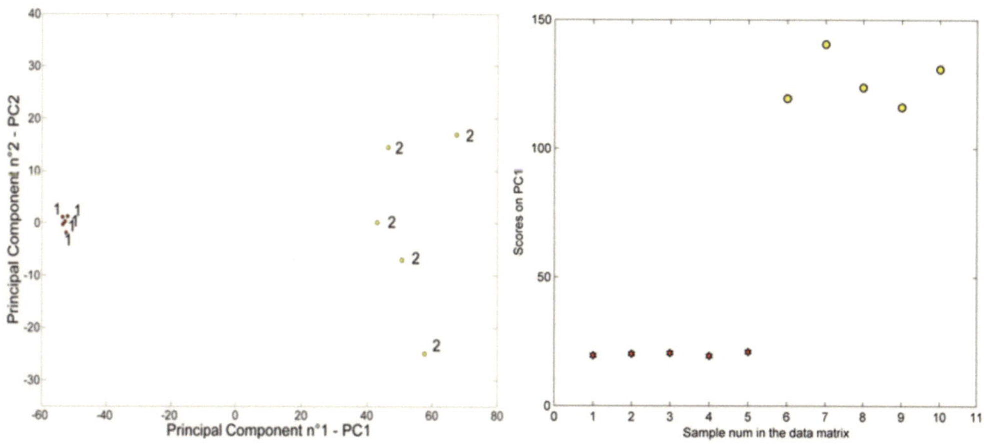

Figure 4. On the left hand side: scores-plot from samples in two first principal components space; **on the right hand side**: samples-scores on PC1.

It was found that the samples are divided into two distinct groups on the initial data. Another way to visualize this division is to represent only the scores on PC1 (on PC2 or PCi) for each sample (see Figure 4). In this case, it is also obvious that the samples 1 to 5 have similar values of PC1 and, thus, form a first group (group 1), while samples 6-10 are the second group (group 2).

Step 5. Calculating factorial contributions - Loadings

A complete interpretation of the results of PCA involves the graph of the loadings, i.e. the projection of the variables in the sample space. But how does one get this? Consider what has been calculated so far: the eigenvalues and eigenvectors of the matrix samples and their factorial coordinates in the space of principal components. We need to calculate the factorial coordinates of variables in the sample space. The results are called "*loadings*" or "*factorial contributions*". You just have to transpose the initial data matrix and repeat the entire calculation. Figure 5 shows the results representing the position of the variables of the problem in the plane formed by the first two PCs.

The graph of the loadings allows us to understand what the characteristic variables of each group of samples on the graph of the scores are. We see, in particular, that the samples of Group 1 are distinguished from the others by variables 420 and 474 while the Group 2 samples are distinguished from the others by variables 520 and 570. In other words, for these two sets of variables, the groups of samples have opposite values in quantitative terms: when a group has high values for the two pairs of variables, then the other group of samples has low values for the same pair of variables, and vice versa.

When the measured variables are of structural (mass spectrometry data) or spectral types (infrared bands or chemical shifts in NMR), the joint interpretation of scores and loadings can be extremely interesting because one can be in a position to know what distinguishes the groups of samples from a molecular point of view.

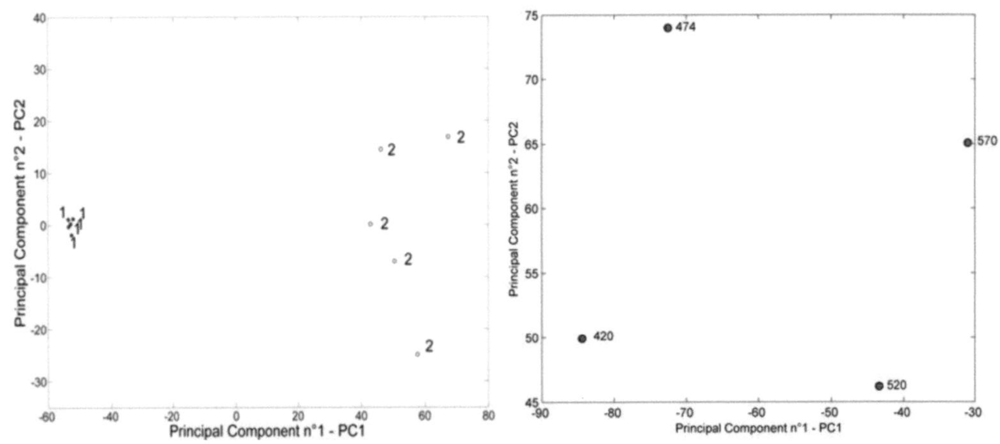

Figure 5. Scores & loadings on the PC1xPC2 plane.

4.2. Research example: Application of PCA on 1H-NMR spectra for the study of the thermal stability of edible oils

Numerous chemical reactions contribute to the chemical and physical aging of oils. During heating, the oils undergo degradation and their functional and organoleptic features are significantly modified. The heating induces chemical reactions such as oxidation, polymerization, hydrolysis and cis/trans isomerization, which have an impact not only on the nutritional value of oils but which may also generate toxic compounds injurious to health [*24*, *25*]. The most important cause of the deterioration of oils is oxidation. Among the oxidation products, our interest has focused especially on secondary oxidation products, such as aldehydes, because they are rarely present in natural unheated oil, as described by Choe et al. [*26*] from the study of secondary oxidation products proposed by Frankel [*27*]. In this work, an analytical approach was first adopted to calculate a new semi-quantitative criterion of the thermal stability of oils. This new test is based on the assumption that by focusing on a selected portion of the 1H-NMR spectra and for a relatively short time, we can model the appearance of aldehydes by a kinetic law of order 1, knowing that the mechanisms actually at work are more complex and associated with radical reactions. In the following pages is presented only that part of this work related to the application of PCA to 1H-NMR data to characterize and follow the effect of temperature and time of heating on the chemical quality of edible oils. For further information about the kinetic study and multiway treatments see [*28*].

4.2.1. Samples

Three types of edible oils were analysed. Rapeseed, sunflower and virgin olive oils were purchased at a local supermarket and used in a thermal oxidation study. Approximately 12 mL of oil was placed in 10 cm diameter glass dishes and subjected to heating in a laboratory oven with temperature control. Each of the three types of oil was heated at 170 °C, 190 °C and 210 °C, each of which is close to home-cooking temperatures. Three samples of 1 g were collected every 30 min until the end of the heating process, fixed at 180 min, resulting in a total of 189 samples to be analysed. Samples were cooled in an ice-water bath for 4 min, in order to stop thermal-oxidative reactions, and then directly analysed.

4.2.2. 1H-NMR spectroscopy

Between 0.3 and 0.5 g of oil was introduced into an NMR tube (I.D. 5 mm) with 700 μL of deuterated chloroform for the sample to reach a filling height of approximately 5 cm. The proton NMR spectrum was acquired at 300.13 MHz on a Bruker 300 Advance Ultrashield spectrometer with a 7.05 T magnetic field. A basic spin echo sequence was applied. The acquisition parameters were: spectral width 6172.8 Hz; pulse angle 90–180°; pulse delay 4.4 μs; relaxation delay 3 s; number of scans 64; plus 2 dummy scans, acquisition time 5.308 s, with a total acquisition time of about 9 min. The experiment was carried out at 25 °C. Spectra were acquired periodically throughout the thermal oxidation process. All plots of 1H NMR spectra or spectral regions were plotted with a fixed value of absolute intensity for comparison.

4.2.3. Data & data processing

The initial matrix (189 samples × 1001 variables) contains 1H-NMR spectra of three oils at three heating temperatures and seven heating times. The computations were performed using the MATLAB environment, version R2007b (Mathworks, Natick, MA, USA).

4.2.4. Results and discussion

The chemical changes taking place in the oils over time and at different temperatures, were monitored by 1H nuclear magnetic resonance spectroscopy. The spectral region between δ 0 and δ 7.2 was not used in this work because the changes observed were not exclusively specific to heating by-products, as were those in the spectral region between δ 9 and δ 10. Figs. 6.1–6.3 present this spectral region for rapeseed oil for different heating times at 170 °C, 190 °C and 210 °C. The increase in the aldehydic protons peaks over time and as a function of temperature is clearly visible. It is to be noted that the temperature has a similar effect in this spectral region for sunflower oil, but with a different patterns of peaks. This reflects differences in the structure of the aldehydes formed in the two oils. The effect of temperature is much smaller for olive oil due to the composition of the fatty acids and a higher concentration of natural antioxidants such as tocopherols.

PCA score plots show classical trajectories with time and temperature of the three oils, and 66% of the total variance is explained by the first 2 components (see Fig. 7). The score plots show that the thermal degradations are quite different, with that of olive oil being much less extensive, with that of rapeseed being more similar to olive oil than sunflower oil. This observation makes sense because olive oil and rapeseed oil have approximately the same range of oleic acid (respectively 55.0–83.0% and 52.0–67.0% [*29*]), which is one of the three major unsaturated fatty acids of edible oils, along with linoleic and linolenic acids, while sunflower oil has a lower content (14.0–38.0%). The smaller spread of scores for olive oil at a given temperature indicates that chemical changes caused by heating over time are more pronounced in rapeseed and sunflower oil than in olive oil. Olive oil is also different from the other two oils in that the orientation of its scores' trajectory is different for each temperature. The plot of the PC1 and PC2 loadings suggests that the chemical evolution of the sunflower oil and olive oil is not characterized by the formation of the same aldehyde protons. The scores for PC1 characterize the time factor for sunflower, rapeseed and olive oils at 170 °C, while the loadings for PC1 characterize the progressive appearance of trans-2-alkenals and E,E-2,4-dialkadienals associated with peaks at δ 9.534, 9.507 and 9.554 (see Fig. 7) and the short chain alkanals associated with the peak at δ 9.766.

The scores for PC2 are related more to the time factor for rapeseed oil and olive oil at 190 and 210 °C. The loadings for PC2 are characterized by several peaks: δ 9.497 and 9.506 (possibly related to the increase of trans-2-alkenals), δ 9.763 and δ 9.786 (N-alkanals and short chain alkanals or oxo-alkanals) and δ 9.553 (possibly 4,5-epoxy trans-2-alkenals). According to Guillén et al. [*30, 31*], these compounds are among the secondary oxidation products. The loadings of each PC give a clear idea of which peaks increase or decrease during the heating and allow a better understanding of what differentiates the oil samples from a chemical point of view. The Euclidean distance in the multivariate space of three first principal components (i.e. those that explain the majority of the variance of the data studied (70%)) allows the proposal of a thermal stability order of studied edible oils. This order is presented in table 7.

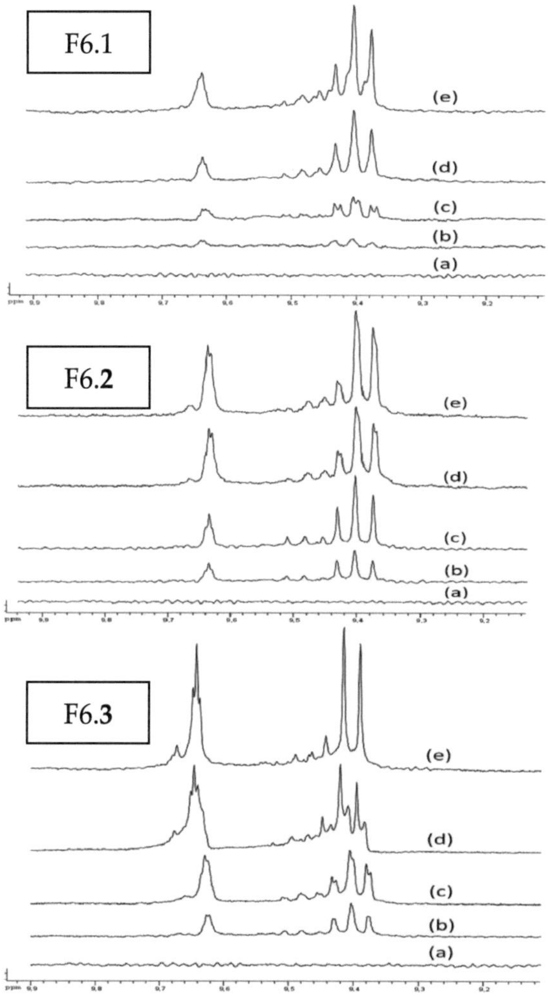

	Heating Conditions (min)
(a)	Not heated
(b)	30
(c)	60
(d)	120
(e)	180

Figure 6. Expanded region between δ 9 and δ 10 of the 1H NMR spectra of rapeseed oil at different times during heating (F5.1) at 170°C, (F5.2) at 190°C, (F5.3) at 210°C.

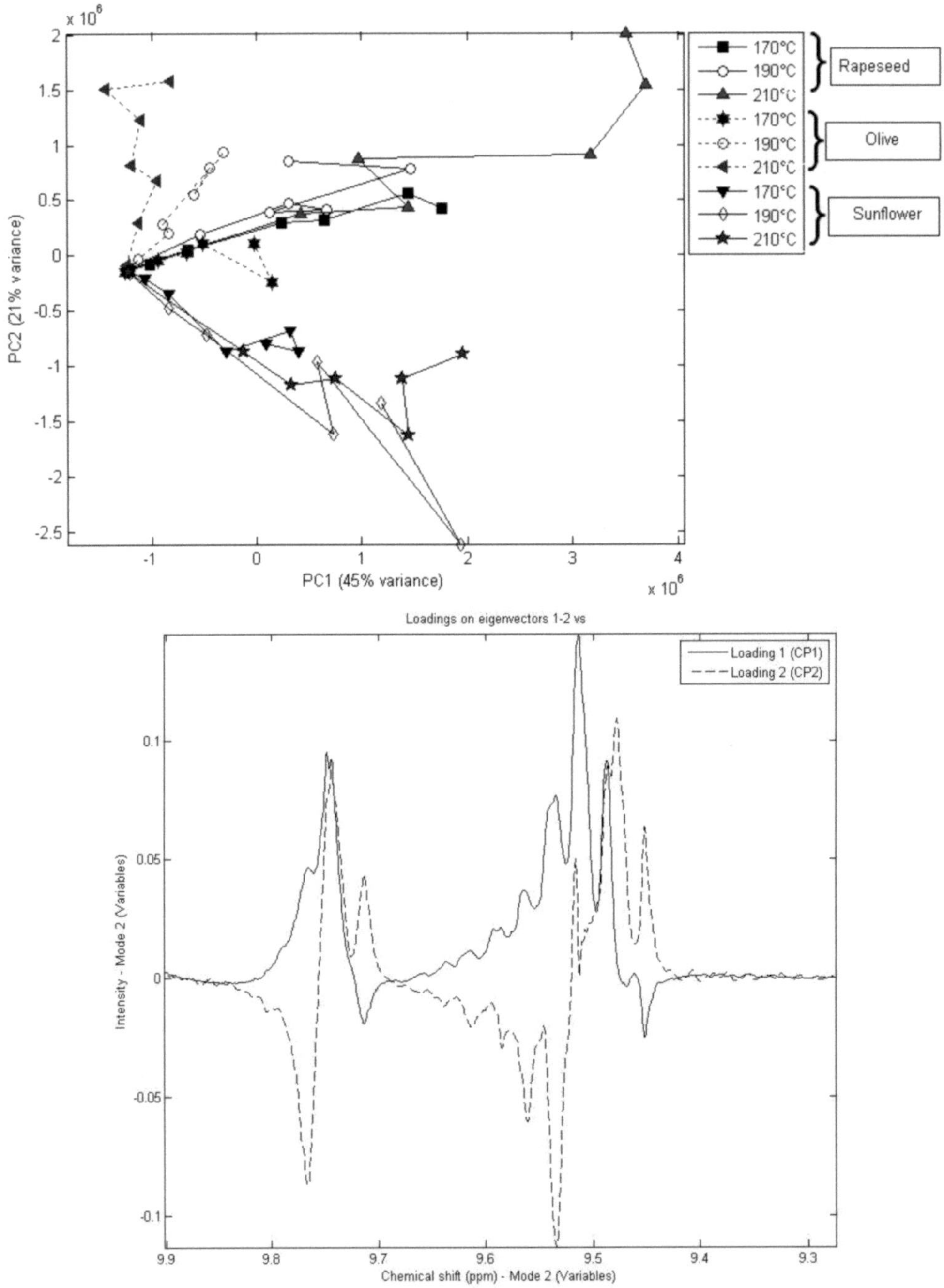

Figure 7. PCA scores (up) and loadings (bottom) of (63 x 1001) data matrix for 3 types of oils and 3 heating temperatures. In both cases, only PC1 and PC2 are represented.

Observed thermal stability order		
Temperature		
190 °C	170 °C	210 °C
OO > SO >= RO	OO > RO >= SO	OO > SO >> RO

Table 7. Thermal stability order of the studied oils given by multivariate distances calculated from coordinates on 3 first PCs. OO: olive oil, RO: rapeseed oil, SO: sunflower oil.

4.2.5. Conclusion

The conclusion of this work indicates that olive oil is more stable than the others (rapeseed and sunflower oil), whatever the heating temperature applied. One reason for this stability is its higher concentration of natural antioxidants, such as tocopherols, and its content of oleic acid with unsaturated bonds, which allow it to participate more easily in the radical oxidation mechanisms which occur during heating and thus protect other fragile compounds in the oil for longer.

This application example of the study of the stability of heated oils shows the power of such tool as PCA in the treatment of multivariate spectroscopic data as used with the characteristic fingerprints of chemical and physical phenomena occurring in the samples over time. The PCA allows the direct use of spectra (or pieces of spectra) or analytical curves instead of the integration values typically used in NMR spectroscopy. The principle presented here is now very common in the field of spectroscopists in general and many monitoring tools or processes for monitoring chemical reactions involve PCA or more robust variants (Robust PCA, RPCA based on robust covariance estimators, RPBA based on projection pursuit, Kernel PCA, etc.).

5. PARAFAC for parallel factor analysis, a generalization of PCA to 3-way data

5.1. Introduction

Usually, physicochemical analysis - or else the more generally parametric monitoring of a number of physicochemical properties of a set of samples - resulted in the construction of a data matrix with samples in rows and physicochemical properties in columns. This data matrix is a mathematical representation of the characteristics of the sample set at time *t* in preparatory and analytical conditions attached to the instant action. Incidentally, the fact of working on a matrix which is a two-dimensional array allows us to speak of two modes or 2-way data. Now, imagine that you reproduce these measurements on several dates $t_1, t_2, ..., t_n$. You no longer have a matrix X (n, p) but the **N** matrices X_i (n, p) of the same size where i is the number of the matrix corresponding to the time t_i. This is known as 3-mode data or three-way data or a "data cube". Figure 8 below illustrates what has been said.

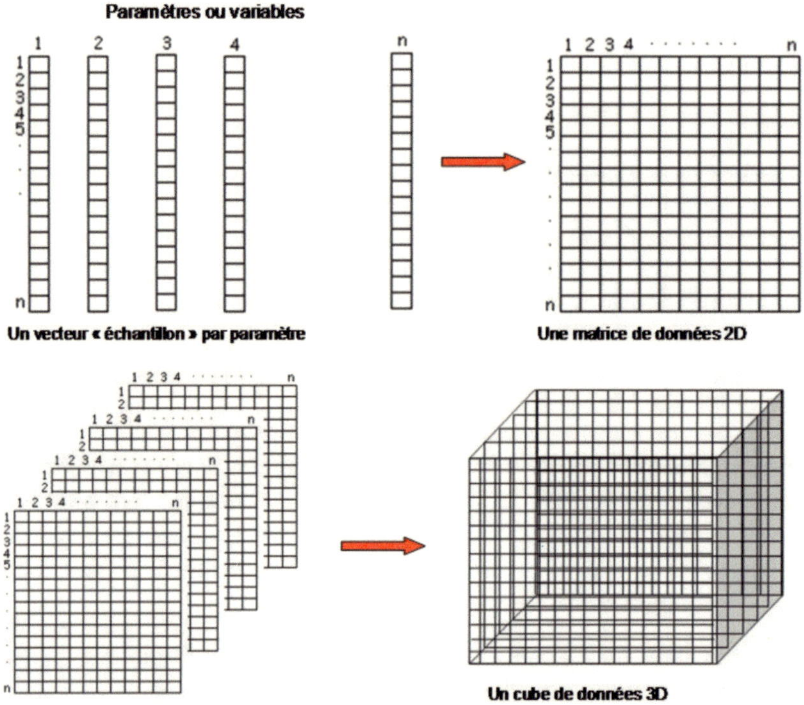

Figure 8. Representation of a matrix and a cube of data from experimental measurements.

At this stage, a key question is raised: how should one analyse all these matrices so as to extract all the relevant information that takes into account the time factor? Several options are available to us. The first would be, for example, an averaging of each parameter per sample measured at the various dates/times. This has the immediate effect of losing the time-based information. By this operation, the effect of time on the measured parameters is no longer observable. We would obtain an average matrix resulting from all of the measured parameters of the study period. The advantage of this in relation to the question is null. Finally, we could consider using the chemometric techniques discussed previously in this book, such as principal component analysis (or others not presented here, such as the hierarchical analysis of each class of matrices stored). But then, it becomes difficult to visualize changes occurring between successive matrices which may correspond to an evolution with time (as in the case of 3D fluorescence spectra which are commonly qualified as second-order data) or to a geographical evolution of the measured parameters (e.g., when studying physicochemical a river port or lagoon in which samples are realized in various places to monitor the level of chemical pollution [32]). The overall interpretation is certainly more difficult.

The solution must be sought to tools capable of taking into account a third or Nth dimension(s) in the data while retaining the ability to account for interactions between factors. The following pages describe two of these tools; probably one of the most famous of them is the PARAllel FACtor model (PARAFAC) introduced independently by Harshman in 1970 [33] and by Carroll & Chang [34] who named the model CANDECOMP (canonical decomposition). The oldest paper we found relating the mathematical idea of PARAFAC was probably published by Hitchcock F.L. in 1927 who presents a method to decompose tensors or polyadics as a sum of products [35]. In this class of tools, the size or dimensions of the dataset are called "modes", hence the name of multi-mode techniques. The most widely used encountered term is "multiway" technique.

5.2. The PARAFAC model

As discussed above, many problems involve chemical 3-ways data tables. Let us come back to the hypothetical example of a liquid chromatography coupled to a fluorescence spectrometer that produces a data set consisting of 3-way "layers" or 2-dimensional matrix sheets coming from the excitation-emission fluorescence spectra collection (Excitation-Emission Matrix: EEM) which vary as a function of elution time.

The fluorescence intensities depend on the excitation and emission wavelengths, and the elution time, these variables will be used to represent the three modes. To exploit these multiway data, the PARAFAC model is an effective possibility to extract useful information. PARAFAC is an important and useful tool for qualitative and quantitative analysis of a set of samples characterized by bilinear data such as EEM in fluorescence spectroscopy. One of the most spectacular examples of the capability of PARAFAC model is the analysis of a mixture of several pure chemical components characterized by bilinear responses. In this case, PARAFAC is able to identify the right pairs of profiles (e.g. emission and excitation profiles) of pure components as well as the right concentration proportions of the mixture. PARAFAC yields unique component solutions. The algorithm is based on a minimizing least squares method. This is to decompose the initial data table in a procedure known as "trilinear decomposition" which gives a unique solution. The trilinear decomposition comes from the model structure and sometimes data itself implies that because of its (their) natural decomposition in 3 modes. The PARAFAC model is a generalization of the PCA itself bilinear, to arrays of higher order (i.e., three or more dimensions). PCA decomposes the data into a two-mode product of a matrix called matrix T scores and a matrix of loadings P describing systematic variations in the data, over a matrix of residues E representing the discrepancies between actual data and the model obtained. Thus, as illustrated by figure 8, the EEM are arranged in a 3-way table (X) of size I x K x J where I is the number of samples, J the number of emission wavelengths, and K the number of excitation wavelengths. Similarly, PARAFAC decomposes X into three matrices (see figure 9): A (scores), B and C (loadings) with the elements a_{if}, b_{jf}, and c_{kf}. In other words, N sets of triads are produced and the trilinear model is usually presented as followed in equation 1 [36]:

$$x_{ijk} = \sum_{f=1}^{F} a_{if} b_{jf} c_{kf} + e_{ijk} \qquad (3)$$
$$(i = 1...I; j = 1...J; k = 1...K)$$

where x_{ijk} is the fluorescence intensity for the ith sample at the emission wavelength j and the excitation wavelength k. The number of columns f in the matrices of loadings is the number of PARAFAC factors and e_{ijk} residues, which account for the variability not represented by the model.

Note the similarities between the PARAFAC model and that of the PCA in schema 2, § "II.A *Some theoretical aspects*". The PARAFAC model is a specific case of the Tucker3 model introduced by Tucker in 1966. The following paragraph presents the essentials about Tucker3 model and proposes some important papers on the theory and applications of this multiway tool. The subsequent paragraph gives a more complete bibliography related to PARAFAC and Tucker3 models on various application areas.

5.3. Tucker3: a generalization of PCA and PARAFAC to higher order

Conceptually, the Tucker3 model [37] is a generalization of two-way data decomposition methods such as PCA or singular value decomposition (SVD) to higher order arrays or tensors [8] and [9]. In such multiway methods, scores and loadings are not distinguishable and are commonly treated as numerically equivalent. Being a generalization of principal component analysis and PARAFAC to multiway data arrays, the Tucker3 model has for its objective to represent the measured data as a linear combination of a small number of optimal, orthogonal factors. For a 3-way data array, the Tucker3 model takes the following form:

$$x_{ijk} = \sum_{u=1}^{r}\sum_{v=1}^{s}\sum_{w=1}^{t} g_{iu} h_{jv} e_{kw} c_{uvw} + \varepsilon_{ijk}$$

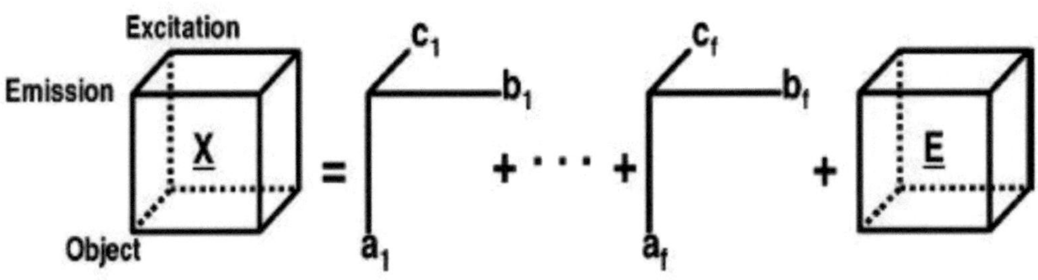

Figure 9. Principle of the decomposition of a 3-way data cube according to the PARAFAC model.

here, x_{ijk} are the measured data, g_{iu}, h_{jv} and e_{kw} are the elements of the loading matrices for each the three ways (with **r**, **s** and **t** factors, respectively) and c_{uvw} are the elements of the core array (of size $\mathbf{r} \times \mathbf{s} \times \mathbf{t}$), while ε_{ijk} are the elements of the array of the residuals. A tutorial on chemical applications of Tucker3 was proposed by Henrion [33], while Kroonenberg [34] gives a detailed mathematical description of the model and discusses advanced issues such as data preparation/scaling and core rotation. For a complete and very pedagogical comparison of Tucker3 with PARAFAC, another multiway procedure, see Jiang [35]. Nevertheless, some aspects of Tucker3 model which distinguish it from PARAFAC have to be discussed here. The first, the Tucker3 model does not impose the extraction of the same number of factors for each mode. Second, the existence of a core array, *C*, governing the interactions between factors allows the modelling of two or more factors that might have the same chromatographic profile but different spectral and/or concentration profiles. Third, the presence of the core cube in the Tucker3 model gets it to appear as a non linear model which is not always appropriate for problems having trilinear structure. But this limitation can be overcome by applying constraints to the core cube *C*. In some cases, with constraints applied to *C* leadings to have only nonzero elements on the superdiagonal of the cube and a number of factors equal on each mode, then the resulting solution is equivalent to the PARAFAC model [*38*].

5.4. Applications: Brief review

Although PARAFAC and Tucker3 as factorial decomposition techniques come from the last century, their routine use in analytical chemistry became popular with the Rasmus Bro's thesis in 1998. Of particular note is the remarkable PhD works published and available on the Department of Food Science website of the Faculty of Life Sciences, University of Copenhagen using the PARAFAC model and many other multivariate and multiway methods in many industrial food sectors[6]. Before the Bro's work and more generally of the Danish team, one can list a number of publications on the application of PARAFAC or Tucker3 in various sciences, but the literature does not appear to show significant, focused and systematic production of this type of model in the world of chemistry. Since the 2000s, multiway techniques have become widespread with strength in analytical chemistry in fields as diverse as food science and food safety, environment, sensory analysis and process chemistry. This explosion of applications in the academic and industrial sectors is linked to the popularization of analytical instruments directly producing multiway data and particularly fluorescence spectrometers for the acquisition matrix of fluorescence which are naturally two-modes data, inherently respecting the notion of tri-linearity and therefore suitable for processing by models such as PARAFAC or Tucker3. Applications of three-way techniques are now too numerous to be cited in their entirely. Therefore, some of the more interesting applications are listed in table 8 below and the reader is encouraged to report himself to reviews included in this table.

---

[6] Department of Food Science, Faculty of Life Sciences, University of Copenhagen, http://www.models.kvl.dk/theses, last visit April, 2012.

Subject	Type	Technique(s)	Year (Ref.)
Foundations of the PARAFAC procedure: Models and conditions		----	1970 [33]
PARAFAC. Tutorial and Applications		General	1997 [39]
Quantification of major metabolites of acetylsalicylic acid		Fluorescence	1998 [40]
Fluorescence of Raw Cane Sugars		Fluorescence	2000 [41]
Determination of chlorophylls and pheopigments		Fluorescence	2001 [42]
Practical aspects of PARAFAC modelling of fluorescence excitation-emission data		Fluorescence	2003 [43]
Evaluation of Two-Dimensional Maps in Proteomics		2D-PAGE imaging	2003 [44]
Evaluation of Processed Cheese During Storage		Front face fluorescence	2003 [45]
Quantification of sulphathiazoles in honeys		Fluorescence	2004 [46]
Evaluation of Light-Induced Oxidation in Cheese		Front face fluorescence	2005 [47]
Olive Oil Characterization		Front face fluorescence	2005 [48]
Detection of Active Photosensitizers in Butter (riboflavine, protoporphyrine, hématoporphyrine, chlorophylle a)		Fluorescence + sensory analysis	2006 [49]
Characterizing the pollution produced by an industrial area in the Lagoon of Venice		Physicochemical analysis	2006 [32]
Calibration of folic acid and methotrexate in human serum samples		Fluorescence	2007 [50]
Quantification of sulphaguanidines in honeys		Fluorescence	2007 [51]
Water distribution in smoked salmon		RMN	2007 [52]
Monitoring of the photodegradation process of polycyclic aromatic hydrocarbons		Fluorescence + HPLC	2007 [53]
Quantification of tetracycline in the presence of quenching matrix effect		Fluorescence	2008 [54]
Fluorescence Spectroscopy: A Rapid Tool for Analyzing Dairy Products		Fluorescence	2008 [55]
Determination of aflatoxin B1 in wheat		Fluorescence	2008 [56]
Study of pesto sauce appearance and of its relation to pigment concentration		Sensory analysis + HPLC-UV	2008 [57]
Determination of vinegar acidity		ATR- IR	2008 [58]
Multi-way models for sensory profiling data		Sensory analysis	2008 [59]
Noodles sensory data analysis		Physicochemical and sensory analysis	2011 [60]
Kinetic study for evaluating the thermal stability of edible oils		1H-NMR	2012 [28]

 Books and thesis;  Scientific papers;  Reviews

Table 8. Bibliography related to PARAFAC and/or Tucker3 models. Theory and applications in areas such as chemicals and food science, medicine and process chemistry.

5.5. A research example: combined utilization of PCA and PARAFAC on 3D fluorescence spectra to study botanical origin of honey

5.5.1. Application of PCA

The interest for the use of chemometric methods to process chromatograms in order to achieve a better discrimination between authentic and adulterated honeys by linear discriminant analysis was demonstrated by our group previously [*61*]. An extent of this work was to quantify adulteration levels by partial least squares analysis [*62*]. This approach was investigated using honey samples adulterated from 10 to 40% with various industrial bee-feeding sugar syrups. Good results were obtained in the characterization of authentic and adulterated samples (96.5% of good classification) using linear discriminant analysis followed by a canonical analysis. This procedure works well but the data acquisition is a bit so long because of chromatographic time scale. A new way for honey analysis was recently investigated with interest: Front-Face Fluorescence Spectroscopy (FFFS). The autofluorescence (intrinsic fluorescence) of the intact biological samples is widely used in biological sciences due to its high sensitivity and specificity. Such an approach increases the speed of analysis considerably and facilitates non-destructive analyses. The non-destructive mode of analysis is of fundamental scientific importance, because it extends the exploratory capabilities to the measurements, allowing for more complex relationships such as the effects of the sample matrix to be assessed or the chemical equilibriums occurring in natural matrices. For a recent and complete review on the use of fluorescence spectroscopy applied on intact food systems see [*63, 64*]. Concerning honey area, FFFS was directly applied on honey samples for the authentication of 11 unifloral and polyfloral honey types [*65*] previously classified using traditional methods such as chemical, pollen, and sensory analysis. Although the proposed method requires significant work to confirm the establishment of chemometric model, the conclusions drawn by the authors are positive about the use of FFFS as a means of characterization of botanical origin of honeys samples. At our best knowledge, the previous mentioned paper is the first work having investigated the potential of 2D-front face fluorescence spectroscopy to determine the botanical origins of honey at specific excitation wavelengths. We complete this work by adopting a 3D approach of measurements. We present here below the first characterization of three clear honey varieties (Acacia, Lavender and Chestnut) by 3D-Front Face Spectroscopy.

5.5.1.1. Samples

This work was carried out on 3 monofloral honeys (Acacia: *Robinia* pseudo-acacia, Lavandula: *Lavandula hybrida* and Chestnut: *Castanea sativa*). Honeys were obtained from French beekeepers. The botanical origin of the samples was certified by quantitative pollen analysis according to the procedure of Louveaux et al. [*66*]. An aliquot part of 10 g of the honey samples was stirred for 10min at low rotation speed (50-80 rpm) after slight warming (40°C, for 1h), allowing the analysis of honeys at room temperature by diminishing potential difficulties due to different crystallization states of samples. Honey samples were pipetted in 3 mL quartz cuvette and spectra were recorded at 20 °C.

5.5.1.2. 3D-Fluorescence spectroscopy

Fluorescence spectra were recorded using a FluoroMax-2 spectrofluorimeter (Spex-Jobin Yvon, Longjumeau, France) mounted with a variable angle front-surface accessory. The incidence angle of the excitation radiation was set at 56° to ensure that reflected light, scattered radiation and depolarisation phenomena were minimised.

The fluorescence excitation spectra were recorded from 280 nm to 550 nm (increment 4 nm; slits: 3 nm, both at excitation and emission), the fluorescence emission spectra were recorded from 280 to 600 nm, respectively. For each sample, three spectra were recorded using different aliquots.

5.5.1.3. Data Processing and Statistical Analysis

The computations were performed using the MATLAB environment, version R2007b [Mathworks, Natick, MA, USA] and with the N-way Toolbox [67]. Each EEM corresponds to a landscape-matrix. For each succession of fluorescence spectra corresponds a collection of matrices which needs to be processed in a specific arrangement. A cube is usually used to organize landscape-matrices as depicted in the figure 10.

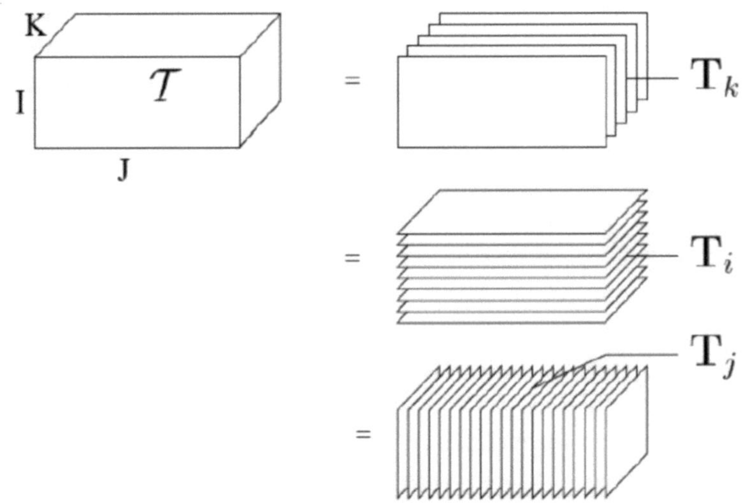

Figure 10. Representation of 3 possible types of arrangements for EEM landscapes processing.

From general point of view, processing this type of data arrangement needs two computing methods: a) 3-way methods such as PARAFAC, Tucker3 or Multiway-PCA and b) 2-way methods such as Principal Component Analysis (PCA) and any other bi-linear technique after data cube unfolding. In the first part of this work, the second approach was used: PCA after the $X^{(1)}$ unfolding of the data cube as illustrated in figure 11. From general point of view, if *I* is the dimension corresponding to samples, it is interesting to make the unfolding of the cube by keeping this dimension unchanged. One can see that $X^{(1)}$ is the unique unfolding that keep the sample dimension unchanged, therefore we took it thereafter for applying PCA.

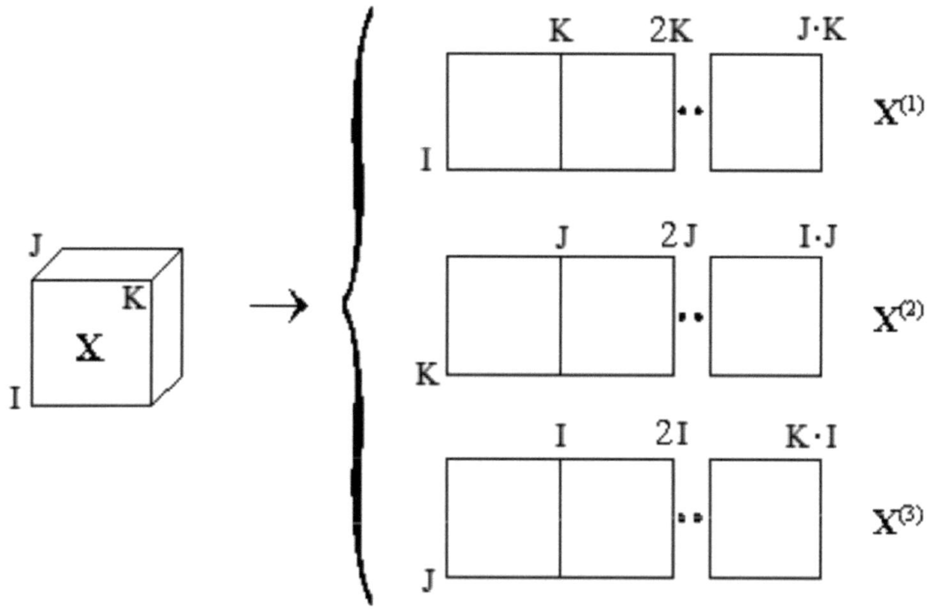

Figure 11. $X^{(1)}$ is obtained by juxtaposition of front slides of X. In the same way, $X^{(2)}$ et $X^{(3)}$ are obtained respectively by juxtaposition of horizontal and lateral slides of X.

5.5.1.4. Results and discussion

Here are below examples of 3D-Front Face Fluorescence spectra of samples of this study. The chemical composition of honey is studied and known to a large extent for nearly 50 years. The presence of compounds beneficial to health for their antioxidant properties or for other reasons is well known, this is the case of polyphenols [*68-74*] or amino acids [*75-83*], some of them are good fluorophores like polyphenols. Works evoke the interest of using these fluorophores as tracers of the floral origin of honeys. For example, the ellagic acid was used as a tracer of heather honey Calluna and Erica species while hesperitin was used to certify citrus honeys [*84, 85*] and abscisic acid was considered as molecular marker of Australian Eucalyptus honeys [*86*]. Kaempferol was used as marker of rosemary honey as well as quercetin for sunflower honey [*87*]. As polyphenols, aromatic amino acids were used to characterize the botanical origin or to test the authenticity of honey, this is the case of phenylalanine and tyrosine for honey lavender [*88*] and the glutamic acid for honeydew honeys [*81*]. Therefore, recording the overall fluorescence spectrum over a large range of wavelengths allows for taking into account the fluorescence emission of the major chemical components described above. In our case, recorded spectra for Acacia, Lavender and Chestnut honeys are similar for certain spectral regions but differ in others proving the existence of different fluorophores. These chemical characteristics specific of the composition are very useful for the distinction of flower varieties. Recording of 3D fluorescence spectra containing all the emission spectra over a wide range of excitation wavelengths can take into account all the information associated with the fluorophores present in honeys.

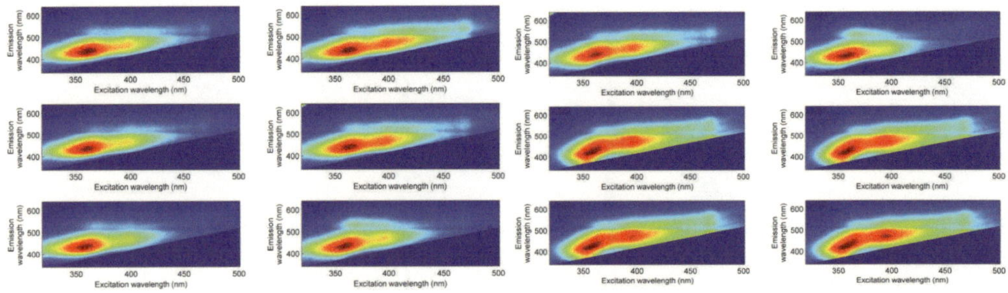

(a)

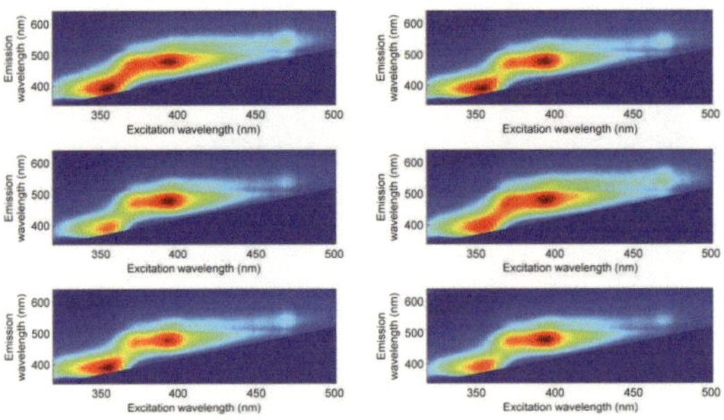

(b)

Figure 12. (a): At left, excitation-emission spectra of Acacia samples. At right, excitation-emission spectra of Lavender samples, (b): Excitation-Emission spectra of Chestnut samples

Figure 13 shows the PC1 vs PC2 scores-plot of the honeys dataset. PCA has played his role of data reduction technique by creating new set of axes. One can visualize all samples simultaneously on a simple xy-graph with only the two first principal components without loose of information. The first obvious conclusion is the natural classes of honeys clearly appear on the scores-plot. PC1 accounts for 58.2% of the total variance while PC2 explains 32.6% of the total variance of the initial data set. The Acacia group is separated from Chestnut and Lavender essentially on PC1, while PC2 is more characteristic of the distinction between Acacia + Lavender and Chestnut. The shape and location of fluorescence islets on 3D-spectra are assets in the distinction of samples as they relate to the chemical compounds that distinguish the groups. It is important to note here that other multivariate techniques exist that could be applied to these data successfully. Particular, the application of the technique ICA (Independent Component Analysis) which is a factorial technique similar to PCA, we would associate with each calculated component a signal having a greater chemical significance. ICA is capable from a mixture of signals to extract mutually independent components explaining more particularly the evolution of a "pure" signal [*89-92*]. In the case presented here, ICA should probably separate more effectively chemical source signals that are causing the observed differences between the honey samples. The chemical interpretation of this result can be assessed using the loadings on PC1 and PC2. As depicted by figure 14, some spectral regions are responsible of these separations observed on the two first axes of PCA. Two excitation/emission maxima (356/376 nm, 468/544 nm) and two excitation/emission minima (350/440 nm, 368/544 nm) are detectable on the PC1 loadings.

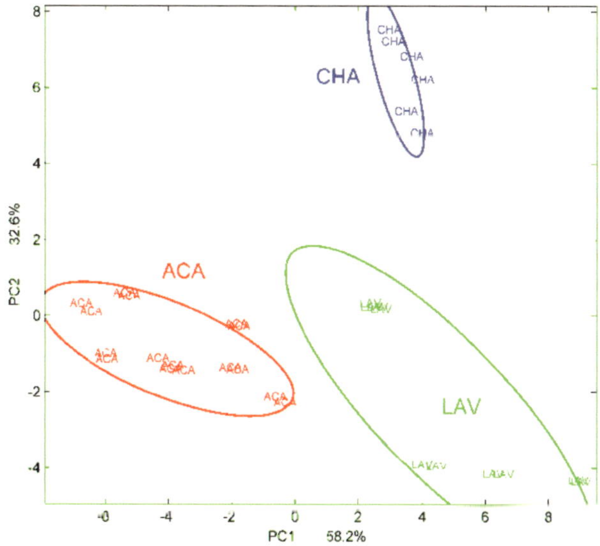

Legend: "ACA = Acacia"; "CHA = Chestnut" ; LAV = Lavender

Figure 13. Ellipsoids of samples distribution drawn with interval confidence at 95%

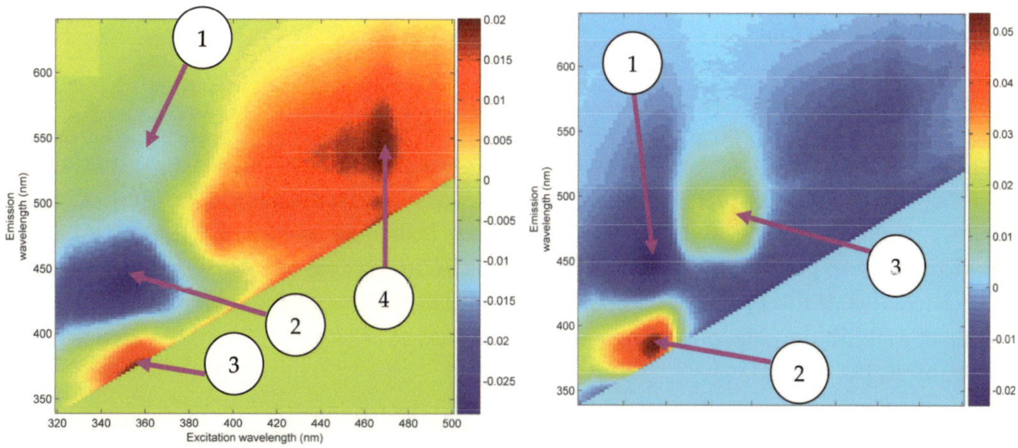

Figure 14. Loadings on PC1 (at left) and PC2 (at right) computed from EEM cube of honey samples after standardization.

In the same manner, the loadings on PC2 show two excitation/emission maxima (356/376 nm, 396/470 nm; respectively zones 2 and 3) and one excitation/emission minima (350/450 nm; zone 1). These maxima and minima are not absolute and cannot be assimilate to true concentrations but represent what spectral zones have largely change in intensity and shape throughout the analysed samples. So, loadings are an overall photography of changes in the samples mainly due to their botanical origins. Figure 15 presents an overview of the main fluorophores potentially present in dairy and food products [63, 93] and some of them are present in honeys too. Here, the chemical interpretation could be made as following. Based on PC1, Acacia samples are mainly distinguished from Lavender and Chestnut honeys by fluorophores appearing in zone 1 and 2, their content in these fluorophores are greater than other samples. Symmetrically, Lavender and Chestnut honeys are more characterized by fluorophores detected in zone 3 and 4. Loadings on PC2 allow a better understanding of what is more characteristic of Chestnut honeys samples compared with others. Zone 2 and 3 are clearly associated with Chestnut samples because the loading values are positive in these regions as the scores are for these samples on the corresponding graph.

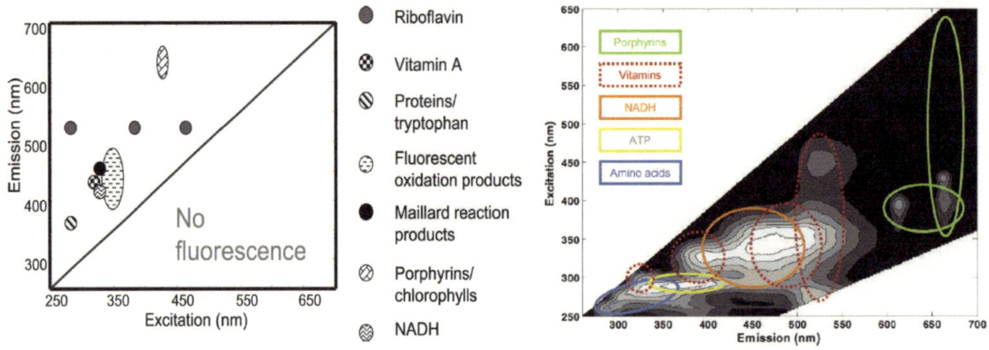

Figure 15. On Left - Excitation and emission maxima of fluorophores present in dairy products (from [93]); On right - Fluorescence landscape map indicating the spectral properties of the selected 11 food-relevant fluorophores (from [63]).

5.5.2. *Application of PARAFAC*

Let us to consider the previous example on the analysis of three varieties of monofloral honeys. The application of the PARAFAC model (with orthogonality constraint on the 3 modes) to the 3-way EEM fluorescence data can take into account the nature of these trilinear data and get factorial cards similarly to PCA. In the case of PARAFAC, it is usual to identify modes of the PARAFAC model from the dimensions of the original data cube. In our case, we built the cube of fluorescence data as *HoneyCube* = [151 x 91 x 32] which is [λ_{em} x λ_{ex} x samples].

Therefore, according to the theoretical model presented above we are able to visualize as many factorial cards as components couples in each mode there are. It is particularly interesting in the case of mode 3 which is mode of samples. The user must specify the number of components in each mode that must be calculated in the model. We built a 3-components model by helping us with the *corcondia* criterion associated with the PARAFAC procedure. The *corcondia* criterion was created and used [94] to facilitate the choice of the optimal number of components in the calculated model. It is a number between 0% (worst model fitting) and 100% (best model fitting). The ideal case for choosing the optimal number of components should be to know the exact number of fluorescence sources in the samples, but in our case this number is unknown. We have consider three sources of fluorescence according to our knowledge of the samples (proteins and free amino acids, NADH from cellular materials and oxidation products + Maillard reaction fluorescent products from heating).The results for mode 1 (emission wavelengths), mode 2 (excitation wavelengths) and mode 3 (samples) are presented below in figures 16 and 17. An interesting parallel between the reconstructed PCA loadings and the PARAFAC loadings in each three modes is possible by forming the good figure arrangement.

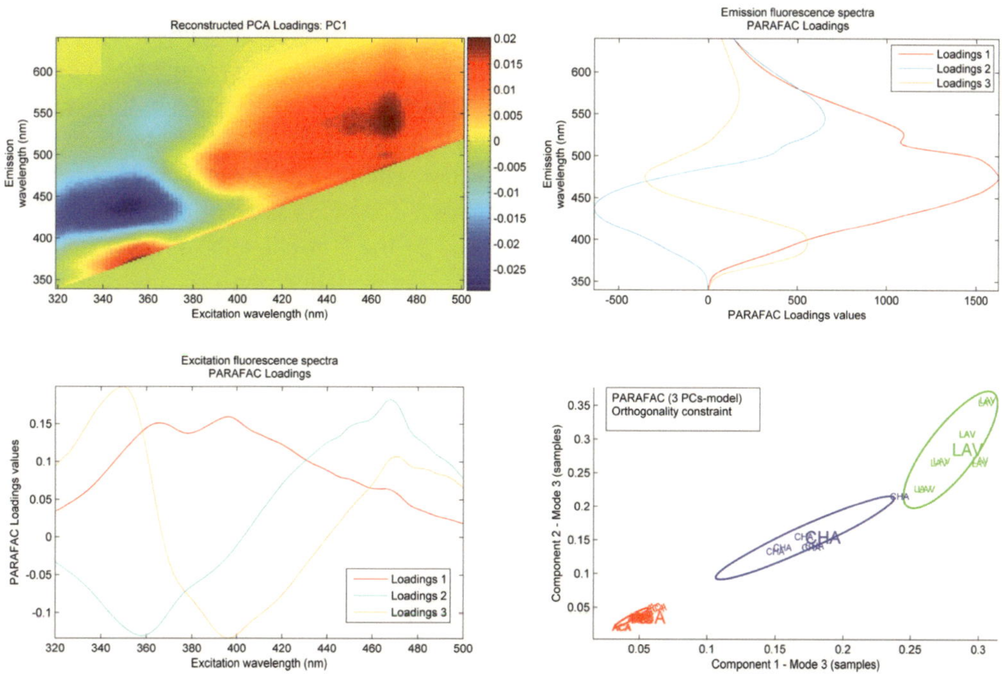

Figure 16. On top left hand side: PCA loadings for component 1; On top right hand side: PARAFAC loadings for Mode 1 (Emission); On bottom left hand side: PARAFAC loadings for Mode 2 (Excitation); On bottom right hand side: PARAFAC loadings for Mode 3 (Samples)

A simultaneous reading of PCA and PARAFAC loadings for honeys EEM matrices gives some elements for understanding what makes distinction between honeys samples on PARAFAC loadings on mode 3. One can see, for example, samples are relatively well separated on the two first PARAFAC loadings in mode 3 (chart on the bottom right). We find good agreement between the PARAFAC profiles and loadings of PCA. Similarly with PCA loadings, those of PARAFAC show the greatest variation of concentration profiles across all samples. And negative areas in the PARAFAC profiles reflect a decrease in fluorescence and therefore a decrease in concentration of the compounds across all the samples. Complementarily, positive PARAFAC profiles indicate an increase in fluorescence and thus the concentration of the corresponding fluorescent compounds in the samples. The first PARAFAC component in mode 3 (chart on the bottom right) allows a good discrimination of the honey samples and the latter is mainly explained by the variation of the fluorescence depicted by the loading 1 (red line) both on the emission (at top and right hand side) and excitation (at bottom and left hand side) graphs. Therefore, Acacia samples have lower content in these compounds than lavender honey samples. Chestnut honeys have an intermediate level of concentration concerning these compounds. Figure 17 gives some elements to more accurately identify the chemical origin of the observed differences between samples on the 2nd PARAFAC component (in blue) on mode 3 (samples). Particularly, it is possible to associated to the 2nd PARAFAC component of mode 3 to Maillard and oxidation products (excitation wavelength: ≈360 nm; emission wavelength: ≈445 nm) depicted by the loadings 2 (blue curve) on the excitation and emission loadings graphs. This observation is explained by the method of harvesting, storage and chemical composition of honey. Indeed, lavender honey experienced a hot extraction and uncapping the honey frames also. This is because the lavender honey crystallizes faster than other honeys because of a higher concentration of sucrose while acacia honey is rich in fructose (sugar does not crystallize) and contains no sucrose. The water content and the glucose / fructose ratio of honeys are also important factors in the crystallization process [95]. The harvesting method is to cause a greater concentration of oxidation products and products of the Maillard reaction.

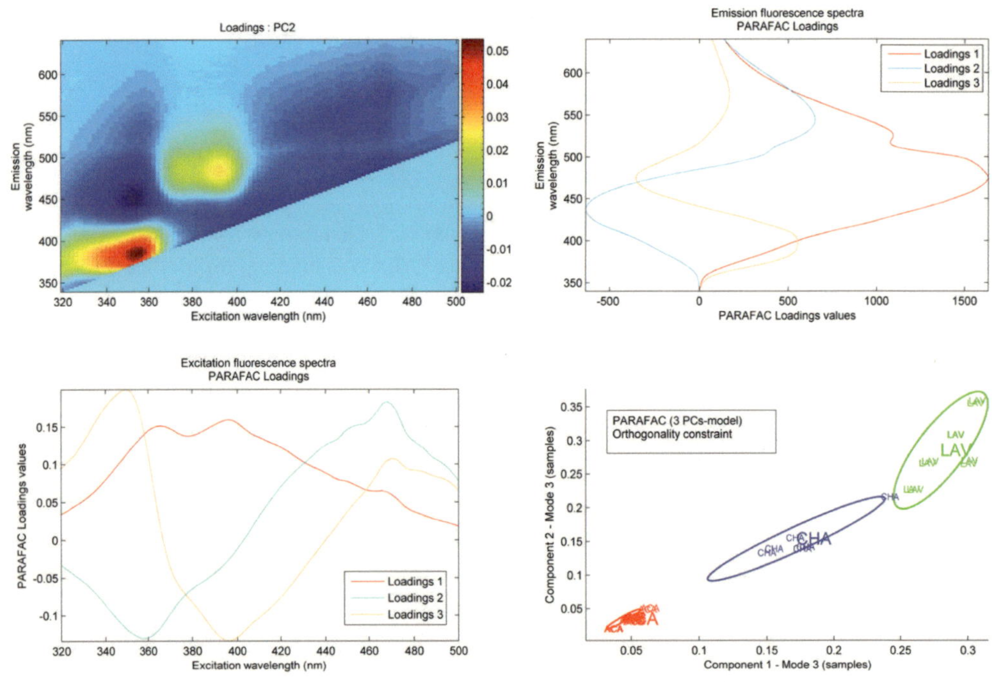

Figure 17. On top left hand side: PCA loadings for component 2; On top right hand side: PARAFAC loadings for Mode 1 (Emission); On bottom left hand side: PARAFAC loadings for Mode 2 (Excitation); On bottom right hand side: PARAFAC loadings for Mode 3 (Samples)

It is clear here that the PARAFAC loadings probably do not reflect pure signals. Indeed, in our case, the appropriate model is probably more complex because we do not know the exact number of fluorophores present in the samples and their contribution is included in each PARAFAC loadings. Thus the chemical interpretation is only partially correct, but nevertheless shows some interesting information on what produces the differences between samples. Another aspect that we have not discussed here is the type of constraint that we have imposed to the model during its construction (orthogonality, non-negativity, unimodality ...). The type of constraints strongly modifies the shape of the PARAFAC loadings obtained and their comparison with PCA is not always possible. We preferred the orthogonality constraint to be in conditions similar to PCA. But this is not necessarily the best choice from spectroscopic point of view, particularly because loadings reflecting concentrations should not be any negative.

5.6. Conclusion

If there were one or two things to note from this application of PARAFAC, it would be, first, the simplicity of the graphical outputs and their complementarities with the results of PCA, and secondly, the possibility offered by the model to process data inherently 3-way or more (without preliminary rearrangement).

In the case of 3D fluorescence data, the loadings of PCA are not easily used directly to interpret and to construct a model of quantitative monitoring, PARAFAC modelling enables this by producing as much as possible loadings representative of changes in concentrations of fluorophores present in the samples, and a good distinction of the studied honey varieties is easily achievable with a simple PARAFAC model.

Appendix

A. Chemometrics books

Here is below a selection of the most interesting books dealing with chemometrics. This list will be convenient as well for the beginner as for the specialist.

- Härdle, W.; Simar, L., *Applied multivariate statistical analysis*. 2nd ed.; Springer: Berlin; New York, 2007; p xii, 458 p.
- Brereton, R. G., *Chemometrics for pattern recognition*. Wiley: Chichester, U.K., 2009; p xvii, 504 p.
- Gemperline, P., *Practical guide to chemometrics*. 2nd ed.; CRC/Taylor & Francis: Boca Raton, 2006; p 541 p.
- Miller, J. N.; Miller, J. C., *Statistics and chemometrics for analytical chemistry*. 5th ed.; Pearson Prentice Hall: Harlow, England; New York, 2005; p xvi, 268 p.
- Beebe, K. R.; Pell, R. J.; Seasholtz, M. B., *Chemometrics : a practical guide*. Wiley: New York, 1998; p xi, 348 p.
- Brown, S. D., Comprehensive chemometrics. Elsevier: Boston, MA, 2009

- Jackson, J. E., *A user's guide to principal components*. Wiley-Interscience: Hoboken, N.J., 2003; p xvii, 569 p.
- Martens, H.; Næs, T., *Multivariate calibration*. Wiley: Chichester [England] ; New York, 1989; p xvii, 419 p.
- Massart, D. L., *Handbook of chemometrics and qualimetrics*. Elsevier: Amsterdam ; New York, 1997.
- Sharaf, M. A.; Illman, D. L.; Kowalski, B. R., *Chemometrics*. Wiley: New York, 1986; p xi, 332 p.

B. Matlab codes

B.1 PCA by decomposition of the covariance matrix

```
% Size of the data matrix X
[n,p]=size(X);
% Column centering
meanX = mean(X);
X_Centred = X - ones(n,1) * meanX;
% Compute variance-covariance matrix
% for scores
CovM_scores = cov(X_Centred);
% for variables
CovM_loadings = cov(X_Centred');
% Compute eigenvalue & eigenvectors by keeping the matrix form
[V_scores,D_scores] = eig(CovM_scores);
D_scores = fliplr(D_scores);D_scores = flipud(D_scores);
V_scores = fliplr(V_scores);

[V_loadings,D_loadings] = eig(CovM_loadings);
D_loadings = fliplr(D_loadings);D_loadings = flipud(D_loadings);
V_loadings = fliplr(V_loadings);
% Compute of total variance of the dataset: TotalVP
TotalVP = trace(D_scores);
% Compute the percentage of variance explained by each principal
%component
```

```
Pourcentages = {};
for i=1:size(D_scores,1)
 Perrcentages{i,1} = strcat('PC',num2str(i));
 Perrcentages{i,2} = num2str(D_scores(i,1)/TotalVP*100);
end;
% Compute factorial coordinates (scores) for each sample point on
% components space
Coord =[];
Coord = X_Cent*V_scores;
% Compute factorial contributions (loadings) for each sample point
on
% components space
Contrib =[];
Contrib = X_Cent'*V_loadings;
```

B.2 PCA by SVD

```
% Size of the data matrix X
[n,p]=size(X);
% Column centering
meanX = mean(X);
X_Centred = X - ones(n,1) * meanX;
% Compute scores and loadings simultaneously
[U SingularValues Loadings] = svd(X_Centred);
% Scores T
T = U * SingularValues;
```

C. Chemometrics software list

- CAMO (http://www.camo.no/) maker of Unscrambler software, for multivariate modelling, prediction, classification, and experimental design.
- Chemometry Consultancy (http://www.chemometry.com/)
- Eigenvector Research (http://www.eigenvector.com/) sells a PLS_Toolbox
- Minitab (http://www.minitab.com/)
- Statcon (http://www.statcon.de/)
- Stat-Ease (http://www.statease.com/) maker of Design-Expert DOE software for experimental design.

- Infometrix, Inc., developer of Pirouette, InStep and LineUp packages for multivariate data analysis. We specialize in integrating chemometrics into analytical instruments and process systems. (http://www.infometrix.com/)
- Umetrics, maker of MODDE software for design of experiments and SIMCA multivariate data analysis. (http://www.umetrics.com/)
- Thermo Electron, offers PLSplus/IQ add-on to GRAMS.
- (http://www.thermo.com/com/cda/landingpage/0,,585,00.html)
- Grabitech Solutions AB, offering MultiSimplex, experimental design & optimization software. (http://www.grabitech.se/)
- PRS Software AS, maker of Sirius software, for data analysis and experimental design. (http://www.prs.no/)
- S-Matrix Corp., maker of CARD, design of experiments software. (http://www.smatrix.com/)
- Applied Chemometrics, source for chemometrics related software, training, and consulting. (http://www.chemometrics.com/)
- DATAN software for multidimensional chemometric analysis. (http://www.multid.se/)

D. Chemometrics websites

- Belgian Chemometrics Society (http://chemometrie.kvcv.be/)
- Département d'Analyse Pharmaceutique et Biomédicale (FABI) à la Vrije Universiteit Brussel (http://www.vub.ac.be/fabi)
- McMaster University Hamilton - Department of Chemical Engineering - Prof. Dora Kourti (http://www.chemeng.mcmaster.ca/faculty/kourti/default.htm)
- NAmICS (http://www.namics.nysaes.cornell.edu/)
- Research NIR Centre of Excellence (NIRCE) - Prof. Paul Geladi (http://www.fieldnirce.org/)
- Université d'Anvers – Groupe de Recherche pour la chimiométrie et l'analyse par rayons x - Prof. Van Espen (http://www.chemometrix.ua.ac.be/)
- Université d'Anvers – Département de Mathématique, Statistique et Actuariat - Prof. Peter Goos (http://www.ua.ac.be/main.aspx?c=peter.goos)
- Université royale vétérinaire et agronomique de Fredriksberg – Groupe de Recherche en Chimiométrie - Prof. Rasmus Bro (http://www.models.kvl.dk/users/rasmus/)
- Université de Louvain - Research Center for Operations Research and Business Statistics - Prof. Martina Vandebroek (http://www.econ.kuleuven.be/martina.vandebroek)
- Université de Nimègue - Chemometrics Research Department - Prof. Lutgarde Buydens (http://www.cac.science.ru.nl/)
- Université de Silésie – Département de chimiométrie - Prof. Beata Walczak (http://chemometria.us.edu.pl/researchBeata.html)
- Johan Trygg (http://www.chemometrics.se/)
- Analytical Chemistry Laboratory, AgroParisTech, France (http://www.chimiometrie.fr)

Author details

Christophe B.Y. Cordella
UMR1145 INRA/AgroParisTech, Institut National de la Recherche Agronomique, Laboratoire de Chimie Analytique, Paris, France

Acknowledgments

I thank Dr Dominique Bertrand (INRA Nantes, France) for proofreading and advice about the PCA and for his availability. I also thank my students who have provided much of the data used in this chapter.

6. References

[1] Leardi, R., Chemometrics: From classical to genetic algorithms. *Grasas Y Aceites* 2002, 53, (1), 115-127.

[2] Siebert, K. J., Chemometrics in brewing - A review. *Journal of the American Society of Brewing Chemists* 2001, 59, (4), 147-156.

[3] Molher-Smith, A.; Nakai, S., Classification of Cheese Varieties by Multivariate Analysis of HPLC Profiles. *Can. Inst. Food Sci. Technol. J.* 1990, 23, (1), 53-58.

[4] Lai, Y. W.; Kemsley, E. K.; Wilson, R. H., Potential of Fourier Transform Infrared Spectroscopy for the Authentication of Vegetable Oils. *J. Agric. Food Chem.* 1994, 42, 1154-1159.

[5] Defernez, M.; Kemsley, E. K.; Wilson, R. H., Use of Infrared Spectroscopy and Chemometrics for the Authentication of Fruit Purees. *Journal of Agricultural and Food Chemistry* 1995, 43, (1), 109-113.

[6] Twomey, M.; Downey, G.; McNulty, P. B., The potential of NIR spectroscopy for the detection of the adulteration of orange juice. *Journal of the Science of Food and Agriculture* 1995, 67, (1), 77-84.

[7] Suchánek, M.; Filipová, H.; Volka, K.; Delgadillo, I.; Davies, A., Qualitative analysis of green coffee by infrared spectrometry. *Fresenius' Journal of Analytical Chemistry* 1996, 354, (3), 327-332.

[8] Baeten, V.; Meurens, M.; Morales, M. T.; Aparicio, R., Detection of Virgin Olive Oil Adulteration by Fourier Transform Raman Spectroscopy. *Journal of Agricultural and Food Chemistry* 1996, 44, (8), 2225-2230.

[9] Downey, G., Food and food ingredient authentication by mid-infrared spectroscopy and chemometrics. *TrAC Trends in Analytical Chemistry* 1998, 17, (7), 418-424.

[10] Garcia-Lopez, C.; Grane-Teruel, N.; Berenguer-Navarro, V.; Garcia-Garcia, J. E.; Martin-Carratala, M. L., Major Fatty Acid Composition of 19 Almond Cultivars of Different Origins. A Chemometric Approach. *Journal of Agricultural and Food Chemistry* 1996, 44, (7), 1751-1755.

[11] Martín Carratalá, M. L.; García-López, C.; Berenguer-Navarro, V.; Grané-Teruel, N., New Contribution to the Chemometric Characterization of Almond Cultivars on the Basis of Their Fatty Acid Profiles. *Journal of Agricultural and Food Chemistry* 1998, 46, (3), 963-967.

[12] Mangas, J. J.; Moreno, J.; Picinelli, A.; Blanco, D., Characterization of Cider Apple Fruits According to Their Degree of Ripening. A Chemometric Approach. *Journal of Agricultural and Food Chemistry* 1998, 46, (10), 4174-4178.

[13] Blanco-Gomis, D.; Mangas Alonso, J. J.; Margolles Cabrales, I.; Arias Abrodo, P., Characterization of Cider Apples on the Basis of Their Fatty Acid Profiles. *Journal of Agricultural and Food Chemistry* 2002, 50, (5), 1097-1100.

[14] König, T.; Schreier, P., Application of multivariate statistical methods to extend the authenticity control of flavour constituents of apple juice. *Zeitschrift für Lebensmitteluntersuchung und -Forschung A* 1999, 208, (2), 130-133.

[15] Ding, H. B.; Xu, R. J., Near-Infrared Spectroscopic Technique for Detection of Beef Hamburger Adulteration. *Journal of Agricultural and Food Chemistry* 2000, 48, (6), 2193-2198.

[16] Maeztu, L.; Andueza, S.; Ibanez, C.; Paz de Pena, M.; Bello, J.; Cid, C., Multivariate Methods for Characterization and Classification of Espresso Coffees from Different Botanical Varieties and Types of Roast by Foam, Taste, and Mouthfeel. *Journal of Agricultural and Food Chemistry* 2001, 49, (10), 4743-4747.

[17] Schoonjans, V.; Massart, D. L., Combining spectroscopic data (MS, IR): exploratory chemometric analysis for characterising similarity/diversity of chemical structures. *Journal of Pharmaceutical and Biomedical Analysis* 2001, 26, (2), 225-239.

[18] Terrab, A.; Vega-Pérez, J. M.; Díez, M. J.; Heredia, F. J., Characterisation of northwest Moroccan honeys by gas chromatographic-mass spectrometric analysis of their sugar components. *Journal of the Science of Food and Agriculture* 2002, 82, (2), 179-185.

[19] Brenna, O. V.; Pagliarini, E., Multivariate Analysis of Antioxidant Power and Polyphenolic Composition in Red Wines. *Journal of Agricultural and Food Chemistry* 2001, 49, (10), 4841-4844.

[20] Destefanis, G.; Barge, M. T.; Brugiapaglia, A.; Tassone, S., The use of principal component analysis (PCA) to characterize beef. *Meat Sci* 2000, 56, (3), 255-9.

[21] Chan, T. F., An Improved Algorithm for Computing the Singular Value Decomposition. *ACM Transactions on Mathematmal Software* 1982, 8, (1), 72-83.

[22] Golub, G.; Kahan, W., Calculating the Singular Values and Pseudo-Inverse of a Matrix. *Journal of the Society for Industrial and Applied Mathematics, Series B: Numerical Analysis* 1965, 2, (2), 205-224.

[23] Mermet, J.-M.; Otto, M.; Widmer, H. M., Multivariate Methods. In *Analytical Chemistry*, Kellner, R., Ed. Wiley-VCH Verlag GmbH: D-69469 Weinheim, Germany, 1998; pp p776-808.

[24] Choe, E.; Min, D. B., Chemistry and reactions of reactive oxygen species in foods. *Crit Rev Food Sci Nutr* 2006, 46, (1), 1-22.

[25] Choe, E.; Min, D. B., Chemistry of deep-fat frying oils. *J Food Sci* 2007, 72, (5), R77-86.

[26] Choe, E.; Min, D. B., Mechanisms and Factors for Edible Oil Oxidation. *Comprehensive Reviews in Food Science and Food Safety* 2006, 5, (4), 169-186.

[27] Frankel, E. N.; Min, D. B.; Smouse, T. H., Chemistry of autoxidation: products and flavor significance. In *Flavor Chemistry of Fats and Oils, Champaign III,* American Oil Chemist's Society: 1985; pp 1-34.

[28] Cordella, C. B.; Tekye, T.; Rutledge, D. N.; Leardi, R., A multiway chemometric and kinetic study for evaluating the thermal stability of edible oils by (1)H NMR analysis: comparison of methods. *Talanta* 2012, 88, 358-68.

[29] Harwood, J. A. R., *Handbook of Olive Oil Analysis and Properties*. Aspen Publishers: Gaithersburg, Maryland, 2000.

[30] Aguila, M. B.; Pinheiro, A. R.; Aquino, J. C.; Gomes, A. P.; Mandarim-de-Lacerda, C. A., Different edible oil beneficial effects (canola oil, fish oil, palm oil, olive oil, and soybean oil) on spontaneously hypertensive rat glomerular enlargement and glomeruli number. *Prostaglandins Other Lipid Mediat* 2005, 76, (1-4), 74-85.

[31] Guillén, M. D.; Goicoechea, E., Oxidation of corn oil at room temperature: Primary and secondary oxidation products and determination of their concentration in the oil liquid matrix from 1H nuclear magnetic resonance data. *Food Chemistry* 2009, 116, (1), 183-192.

[32] Carrer, S.; Leardi, R., Characterizing the pollution produced by an industrial area: chemometric methods applied to the Lagoon of Venice. *Sci Total Environ* 2006, 370, (1), 99-116.

[33] Harshman, R. A., Foundations of the PARAFAC procedure: Models and conditions for an "explanatory" multimodal factor analysis.(University Microfilms, Ann Arbor, Michigan, No. 10,085). *UCLA Working Papers in Phonetics* 1970, 16, 1-84.

[34] Carroll, J.; Chang, J.-J., Analysis of individual differences in multidimensional scaling via an n-way generalization of "Eckart-Young" decomposition. *Psychometrika* 1970, 35, (3), 283-319.

[35] Hitchcock, F. L., The expression of a tensor or a polyadic as a sum of products. *Journal of Mathematics and Physics* 1927, 6, 164-189.

[36] Booksh, K. S.; Kowalski, B. R., Calibration method choice by comparison of model basis functions to the theoretical instrumental response function. *Analytica Chimica Acta* 1997, 348, 1-9.

[37] Tucker, L. R., Some mathematical notes on three-mode factor analysis. *Psychometrika* 1966, 31, (3), 279-311.

[38] Kiers, H., Hierarchical relations among three-way methods. *Psychometrika* 1991, 56, (3), 449-470.

[39] Bro, R., PARAFAC. Tutorial and applications. *Chemometrics and Intelligent Laboratory Systems* 1997, 38, (2), 149-171.

[40] Esteves da Silva, C. G. J.; Novais, A. G. S., Trilinear PARAFAC decomposition of synchronous fluorescence spectra of mixtures of themajor metabolites of acetylsalicylic acid. *Analyst* 1998, 123, (10), 2067-2070.

[41] Baunsgaard, D.; Norgaard, L.; Godshall, M. A., Fluorescence of Raw Cane Sugars Evaluated by Chemometrics. *Journal of Agricultural and Food Chemistry* 2000, 48, (10), 4955-4962.

[42] Moberg, L.; Robertsson, G.; Karlberg, B., Spectrofluorimetric determination of chlorophylls and pheopigments using parallel factor analysis. *Talanta* 2001, 54, (1), 161-70.

[43] Andersen, C. M.; Bro, R., Practical aspects of PARAFAC modeling of fluorescence excitation-emission data. *Journal of Chemometrics* 2003, 17, (4), 200-215.

[44] Marengo, E.; Leardi, R.; Robotti, E.; Righetti, P. G.; Antonucci, F.; Cecconi, D., Application of Three-Way Principal Component Analysis to the Evaluation of Two-Dimensional Maps in Proteomics. *Journal of Proteome Research* 2003, 2, (4), 351-360.

[45] Christensen, J.; Povlsen, V. T.; SÃ¸rensen, J., Application of Fluorescence Spectroscopy and Chemometrics in the Evaluation of Processed Cheese During Storage. *Journal of Dairy Science* 2003, 86, (4), 1101-1107.

[46] Mahedero, M. C.; Diaz, N. M.; Muñoz de la Peña, A.; Espinosa Mansilla, A.; Gonzalez Gomez, D.; Bohoyo Gil, D., Strategies for solving matrix effects in the analysis of sulfathiazole in honey samples using three-way photochemically induced fluorescence data. *Talanta* 2005, 65, (3), 806-813.

[47] Andersen, C. M.; Vishart, M.; Holm, V. K., Application of Fluorescence Spectroscopy in the Evaluation of Light-Induced Oxidation in Cheese. *Journal of Agricultural and Food Chemistry* 2005, 53, (26), 9985-9992.

[48] Guimet, F.; Ferre, J.; Boque, R.; Vidal, M.; Garcia, J., Excitation-Emission Fluorescence Spectroscopy Combined with Three-Way Methods of Analysis as a Complementary Technique for Olive Oil Characterization. *Journal of Agricultural and Food Chemistry* 2005, 53, (24), 9319-9328.

[49] Wold, J. P.; Bro, R.; Veberg, A.; Lundby, F.; Nilsen, A. N.; Moan, J., Active Photosensitizers in Butter Detected by Fluorescence Spectroscopy and Multivariate Curve Resolution. *Journal of Agricultural and Food Chemistry* 2006, 54, (26), 10197-10204.

[50] Muñoz de la Peña, A.; Durán Merás, I.; Jiménez Girón, A.; Goicoechea, H. C., Evaluation of unfolded-partial least-squares coupled to residual trilinearization for four-way calibration of folic acid and methotrexate in human serum samples. *Talanta* 2007, 72, (4), 1261-1268.

[51] Muñoz de la Peña, A.; Mora Diez, N.; Mahedero García, M. C.; Bohoyo Gil, D.; Cañada-Cañada, F., A chemometric sensor for determining sulphaguanidine residues in honey samples. *Talanta* 2007, 73, (2), 304-313.

[52] Løje, H.; Green-Petersen, D.; Nielsen, J.; Jørgensen, B. M.; Jensen, K. N., Water distribution in smoked salmon. *Journal of the Science of Food and Agriculture* 2007, 87, (2), 212-217.

[53] Bosco, M. V.; Larrechi, M. S., PARAFAC and MCR-ALS applied to the quantitative monitoring of the photodegradation process of polycyclic aromatic hydrocarbons using three-dimensional excitation emission fluorescent spectra: Comparative results with HPLC. *Talanta* 2007, 71, (4), 1703-1709.

[54] Rodriguez, N.; Ortiz, M. C.; Sarabia, L. A., Fluorescence quantification of tetracycline in the presence of quenching matrix effect by means of a four-way model. *Talanta* 2009, 77, (3), 1129-1136.

[55] Andersen, C. M.; Mortensen, G., Fluorescence Spectroscopy: A Rapid Tool for Analyzing Dairy Products. *Journal of Agricultural and Food Chemistry* 2008, 56, (3), 720-729.

[56] Hashemi, J.; Kram, G. A.; Alizadeh, N., Enhanced spectrofluorimetric determination of aflatoxin B1 in wheat by second-order standard addition method. *Talanta* 2008, 75, (4), 1075-81.

[57] Masino, F.; Foca, G.; Ulrici, A.; Arru, L.; Antonelli, A., A chemometric study of pesto sauce appearance and of its relation to pigment concentration. *Journal of the Science of Food and Agriculture* 2008, 88, (8), 1335-1343.

[58] Moros, J.; Iñón, F. A.; Garrigues, S.; de la Guardia, M., Determination of vinegar acidity by attenuated total reflectance infrared measurements through the use of second-order absorbance-pH matrices and parallel factor analysis. *Talanta* 2008, 74, (4), 632-641.

[59] Bro, R.; Qannari, E. M.; Kiers, H. A. L.; Næs, T.; Frøst, M. B., Multi-way models for sensory profiling data. *Journal of Chemometrics* 2008, 22, (1), 36-45.

[60] Cordella, C. B. Y.; Leardi, R.; Rutledge, D. N., Three-way principal component analysis applied to noodles sensory data analysis. *Chemometrics and Intelligent Laboratory Systems* 2011, 106, (1), 125-130.

[61] Cordella, C. B.; Militao, J. S.; Cabrol-Bass, D., A simple method for automated pretreatment of usable chromatographic profiles in pattern-recognition procedures: application to HPAEC-PAD chromatograms of honeys. *Anal Bioanal Chem* 2003, 377, (1), 214-9.

[62] Cordella, C. B.; Militao, J. S.; Clement, M. C.; Cabrol-Bass, D., Honey characterization and adulteration detection by pattern recognition applied on HPAEC-PAD profiles. 1. Honey floral species characterization. *J Agric Food Chem* 2003, 51, (11), 3234-42.

[63] Christensen, J.; Norgaard, L.; Bro, R.; Engelsen, S. B., Multivariate autofluorescence of intact food systems. *Chem Rev* 2006, 106, (6), 1979-94.

[64] Sádecká, J.; Tóthová, J., Fluorescence spectroscopy and chemometrics in the food classification – a review. *Czech J. Food Sci.* 2007, 25, 159-174.

[65] Ruoff, K.; Luginbuhl, W.; Kunzli, R.; Bogdanov, S.; Bosset, J. O.; von der Ohe, K.; von der Ohe, W.; Amado, R., Authentication of the botanical and geographical origin of

honey by front-face fluorescence spectroscopy. *J Agric Food Chem* 2006, 54, (18), 6858-66.

[66] Louveaux J.; Maurizio A.; G., V., Methods of Melissopalynology. *Bee World* 1978, 59, 139-157.

[67] Andersson, C. A.; Bro, R., The N-way Toolbox for MATLAB. *Chemometrics and Intelligent Laboratory Systems* 2000, 52, (1), 1-4.

[68] Amiot, M. J.; Aubert, S.; Gonnet, M.; Tacchini, M., Les composés phénoliques des miels: étude préliminaire sur l'identification et la quantification par familles. *Apidologie* 1989, 20, 115-125.

[69] Andrade, P.; Ferreres, F.; Amaral, M. T., Analysis of Honey Phenolic Acids by HPLC, Its Application to Honey Botanical Characterization. *Journal of Liquid Chromatography & Related Technologies* 1997, 20, (14), 2281-2288.

[70] Martos, I.; Ferreres, F.; Tomas-Barberan, F. A., Identification of flavonoid markers for the botanical origin of Eucalyptus honey. *J Agric Food Chem* 2000, 48, (5), 1498-502.

[71] Martos, I.; Ferreres, F.; Yao, L.; D'Arcy, B.; Caffin, N.; Tomas-Barberan, F. A., Flavonoids in monospecific eucalyptus honeys from Australia. *J Agric Food Chem* 2000, 48, (10), 4744-8.

[72] Yao, L.; Datta, N.; TomÃ¡s-BarberÃ¡n, F. A.; Ferreres, F.; Martos, I.; Singanusong, R., Flavonoids, phenolic acids and abscisic acid in Australian and New Zealand Leptospermum honeys. *Food Chemistry* 2003, 81, (2), 159-168.

[73] Yao, L.; Jiang, Y.; D'Arcy, B.; Singanusong, R.; Datta, N.; Caffin, N.; Raymont, K., Quantitative high-performance liquid chromatography analyses of flavonoids in Australian Eucalyptus honeys. *J Agric Food Chem* 2004, 52, (2), 210-4.

[74] Yao, L.; Jiang, Y.; Singanusong, R.; D'Arcy, B.; Datta, N.; Caffin, N.; Raymont, K., Flavonoids in Australian Melaleuca, Guioa, Lophostemon, Banksia and Helianthus honeys and their potential for floral authentication. *Food Research International* 2004, 37, (2), 166-174.

[75] Wang, J.; Li, Q. X.; Steve, L. T., Chapter 3 - Chemical Composition, Characterization, and Differentiation of Honey Botanical and Geographical Origins. In *Advances in Food and Nutrition Research*, Academic Press: Vol. Volume 62, pp 89-137.

[76] Senyuva, H. Z.; Gilbert, J.; Silici, S.; Charlton, A.; Dal, C.; Gurel, N.; Cimen, D., Profiling Turkish honeys to determine authenticity using physical and chemical characteristics. *J Agric Food Chem* 2009, 57, (9), 3911-9.

[77] Rebane, R.; Herodes, K., Evaluation of the botanical origin of Estonian uni- and polyfloral honeys by amino acid content. *J Agric Food Chem* 2008, 56, (22), 10716-20.

[78] Perez, R. A.; Iglesias, M. T.; Pueyo, E.; Gonzalez, M.; de Lorenzo, C., Amino acid composition and antioxidant capacity of Spanish honeys. *J Agric Food Chem* 2007, 55, (2), 360-5.

[79] Iglesias, M. T.; Martin-Alvarez, P. J.; Polo, M. C.; de Lorenzo, C.; Gonzalez, M.; Pueyo, E., Changes in the free amino acid contents of honeys during storage at ambient temperature. *J Agric Food Chem* 2006, 54, (24), 9099-104.

[80] Iglesias, M. T.; Martin-Alvarez, P. J.; Polo, M. C.; de Lorenzo, C.; Pueyo, E., Protein analysis of honeys by fast protein liquid chromatography: application to differentiate floral and honeydew honeys. *J Agric Food Chem* 2006, 54, (21), 8322-7.

[81] Iglesias, M. T.; De Lorenzo, C.; Del Carmen Polo, M.; Martin-Alvarez, P. J.; Pueyo, E., Usefulness of amino acid composition to discriminate between honeydew and floral honeys. Application to honeys from a small geographic area. *J Agric Food Chem* 2004, 52, (1), 84-9.

[82] Bergner, K. G.; Hahn, H., [Phenylalanine content of honeys]. *Z Ernahrungswiss* 1972, 11, (1), 47-54.

[83] Pirini, A.; Conte, L. S.; Francioso, O.; Lercker, G., Capillary gas chromatographic determination of free amino acids in honey as a means of discrimination between different botanical sources. *Journal of High Resolution Chromatography* 1992, 15, (3), 165-170.

[84] Ferreres, F.; García-Viguera, C.; Tomás-Lorente, F.; Tomás-Barberán, F. A., Hesperetin: A marker of the floral origin of citrus honey. *Journal of the Science of Food and Agriculture* 1993, 61, (1), 121-123.

[85] Tomás-Barberán, F. A.; Martos, I.; Ferreres, F.; Radovic, B. S.; Anklam, E., HPLC flavonoid profiles as markers for the botanical origin of European unifloral honeys. *Journal of the Science of Food and Agriculture* 2001, 81, (5), 485-496.

[86] Yao, L.; Jiang, Y.; Singanusong, R.; Datta, N.; Raymont, K., Phenolic acids and abscisic acid in Australian Eucalyptus honeys and their potential for floral authentication. *Food Chemistry* 2004, 86, (2), 169-177.

[87] Pyrzynska, K.; Biesaga, M., Analysis of phenolic acids and flavonoids in honey. *TrAC Trends in Analytical Chemistry* 2009, 28, (7), 893-902.

[88] Bouseta, A.; Scheirman, V.; Collin, S., Flavor and Free Amino Acid Composition of Lavender and Eucalyptus Honeys. *Journal of Food Science* 1996, 61, (4), 683-687.

[89] Comon, P., Independent component analysis, A new concept? *Signal Processing* 1994, 36, (3), 287-314.

[90] Bouveresse, D. J.; Benabid, H.; Rutledge, D. N., Independent component analysis as a pretreatment method for parallel factor analysis to eliminate artefacts from multiway data. *Anal Chim Acta* 2007, 589, (2), 216-24.

[91] Mehrabian, H.; Lindvere, L.; Stefanovic, B.; Martel, A. L., A constrained independent component analysis technique for artery-vein separation of two-photon laser scanning microscopy images of the cerebral microvasculature. *Med Image Anal* 2012, 16, (1), 239-51.

[92] Liu, X.; Liu, F.; Zhang, Y.; Bai, J., Unmixing dynamic fluorescence diffuse optical tomography images with independent component analysis. *IEEE Trans Med Imaging* 2011, 30, (9), 1591-604.

[93] Andersen, C. M. l.; Mortensen, G., Fluorescence Spectroscopy: A Rapid Tool for Analyzing Dairy Products. *Journal of Agricultural and Food Chemistry* 2008, 56, (3), 720-729.

[94] Bro, R.; Kiers, H. A. L., A new efficient method for determining the number of components in PARAFAC models. *Journal of Chemometrics* 2003, 17, (5), 274-286.

[95] White, J. W., *Composition of American honeys*. U. S. Govt. Print. Off.: Washington,, 1962; p iv, 124 p.

Chapter 2

Mineralogy and Geochemistry of Sub-Bituminous Coal and Its Combustion Products from Mpumalanga Province, South Africa

S. A. Akinyemi, W. M. Gitari, A. Akinlua and L. F. Petrik

Additional information is available at the end of the chapter

http://dx.doi.org/10.5772/50692

1. Introduction

Coal forms from the accumulation and physical and chemical alteration of plants remains that settle in swampy areas and form peat, which thickens until heat and pressure transform it into the coal we use. The coal we use is combustible sedimentary rock composed of carbon, hydrogen, oxygen, nitrogen, sulphur, and various trace elements (it has a carbonaceous content of more than 50 % by weight and more than 70 % by volume). As much as 70 % of the South African estimated coal reserve is located in the Waterberg, Witbank, and Highveld coalfields, as well as lesser amounts in the Ermelo, Free State and Springbok Flats coalfields. However, the Witbank and Highveld coalfields are approaching exhaustion (estimated 9 billion tons of recoverable coal remaining in each), while the coal quality or mining conditions in the Waterberg, Free State and Springbok Flats coalfields are significant barriers to immediate, conventional exploitation [1]. South Africa is the third largest coal producer in the world, and coal accounts for 64 % of South Africa's primary energy supply [2]. Electricity generation accounts for 61 % of the total coal consumption in South Africa and more than 90 % of the country's electricity requirements are provided for by coal-fired power plants [2]. South African coals are generally low in sulphur, nitrogen and phosphorus, and in the case of the first two the contents are dependent on maceral composition and rank [3, 4].

The inorganic elements in coal can have profound environmental, economic, technological and human health impacts [5, 6]. Consequently, knowledge of their concentration is necessary when evaluating coals for combustion and conversion and also to evaluate potential negative environmental and health impacts resulting from coal use. Trace elements in coal are emitted into flue gas, fly ash or bottom ash of combustion plants. In a flue gas stream, trace elements are fixed in ash particles and in by-products such as gypsum and sludge if wet flue gas desulphurization unit is equipped.

Coal fly ash is a solid residue from the combustion processes of pulverised coal for the production of electrical power in power generating stations, especially when low-grade coal is burnt to generate electricity [7, 8, 9]. The coal burning power stations in the Mpumalanga Province, South Africa generates over 36.7 Mt of fly ash annually in which only 5 % is currently utilized, the rest being disposed of in the ash dams, landfills, or ponds [9, 10]. During combustion, mineral impurities in the coal, such as clay, feldspars, and quartz, are fused in suspension and float out of the combustion chamber with exhaust gases. As the fused material rises, it cools and solidifies into spherical glassy particles called fly ash [11]. The properties of the coal fly ash depend on the physical and chemical properties of the parent coal, coal particle size, the burning process and the type of ash collector.

This article presents results obtained from mineralogical and chemical characterization of coal and its combustion products from a coal burning power station in the Mpumalanga Province, South Africa. The main objective of the study is to understand the role of combustion process, chemical interaction of fly ash with ingressed CO_2 and percolating rain water on the mineralogy and chemical compositions of fly ash.

2. Methodology/research approach

2.1. XRF and LA-ICPMS analyses

Pulverised coal samples and its combustion products were analysed for major element using Axios instrument (PANalytical) with a 2.4 kWatt Rh X-ray Tube. Further, the same set of samples were analysed for trace element using LA-ICPMS instrumental analysis. LA-ICP-MS is a powerful and sensitive analytical technique for multi-elemental analysis. The laser was used to vaporize the surface of the solid sample and it was the vapour, and any particles, which was then transported by the carrier gas flow to the ICP-MS. The detailed procedures for sample preparation for both analytical techniques are reported below.

2.1.1. Fusion bead method for Major element analysis

- Weigh 1.0000 g ± 0.0009 g of milled sample
- Place in oven at 110 ºC for 1 hour to determine H_2O^-
- Place in oven at 1000 ºC for 1 hour to determine LOI
- Add 10.0000 g ± 0.0009 g Claisse flux and fuse in M4 Claisse fluxer for 23 minutes.
- 0.2 g of $NaCO_3$ was added to the mix and the sample+flux+$NaCO_3$ was pre-oxidized at 700 °C before fusion.
- Flux type: Ultrapure Fused Anhydrous Li-Tetraborate-Li-Metaborate flux (66.67 % $Li_2B_4O_7$ + 32.83 % $LiBO_2$) and a releasing agent Li-Iodide (0.5 % LiI).

2.1.2. Pressed pellet method for Trace element analysis

- Weigh 8 g ± 0.05 g of milled powder
- Mix thoroughly with 3 drops of Mowiol wax binder
- Press pellet with pill press to 15 ton pressure
- Dry in oven at 100 ºC for half an hour before analysing.

2.2. Loss on ignition determination

Loss on Ignition (LOI) is a test used in XRF major element analysis which consists of strongly heating a sample of the material at a specified temperature, allowing volatile substances to escape or oxygen is added, until its mass ceases to change. The L.O.I. is made of contributions from the volatile compounds H_2O^+, OH^-, CO_2, F^-, Cl^-, S; in parts also K^+ and Na^+ (if heated for too long); or alternatively added compounds O_2 (oxidation, e.g. FeO to Fe_2O_3), later CO_2 (CaO to $CaCO_3$). In pyro-processing and the mineral industries such as lime, calcined bauxite, refractories or cement manufacture, the loss on ignition of the raw material is roughly equivalent to the loss in mass that it will undergo in a kiln, furnace or smelter.

2.3. XRD analysis

Coal samples and its combustion products were analysed for mineralogical composition by X-ray diffraction (XRD). A Philips PANalytical instrument with a pw 3830 X-ray generator operated at 40 kV and 25 mA was used. The pulverised samples were oven-dried at 100 °C for 12 h to remove the adsorbed water. The samples were pressed into rectangular aluminium sample holders using an alcohol wiped spatula and then clipped into the instrument sample holder. The samples were step-scanned from 5 to 85 degrees 2 theta scale at intervals of 0.02 and counted for 0.5 sec per step.

2.4. Scanning Electron Microscopy and Electron Dispersive X-ray Spectroscopy (SEM/EDS)

Microstructural and chemical composition investigations of coal and coal ash were carried out by scanning electron microscopy/electron dispersive x-ray spectroscopy (SEM/EDS). For SEM/EDS aluminium stubs were coated with carbon glue; when the glue was dry, but still sticky; a small amount of powder residue samples was sprinkled onto the stub. The excess residue sample powder was tapped off and the glue allowed complete drying. The residue samples were then coated with carbon in an evaporation coater and were ready for analysis with the SEM. The SEM is an FEI Nova NanoSEM (Model: Nova NanoSEM 230); The EDS analyses were determined at 20 Kv and 5 mm working distance. The EDS detector is an Oxford X-Max (large area silicon drift detector) using the software program INCA-(INCAmicaF+ electronics and INCA Feature particle analysis software).

2.5. Transmission Electron Microscopy (TEM)

Pulverized sample (~1-2) g of the coal and coal ash samples was poured into a small conical container while little amount of ethanol was added to the sample to serve as medium for solution. This solution was then placed inside the centrifuge for few minutes (5 mins), drops of the stirred solution was then placed on a labelled 200 μm and 400 μm of copper grid underlain by a filter paper with a hot lamp light focused directly on the samples to dry up the earlier added ethanol. The resultant mixture was placed inside the air gun channel so as to project the beam on it for image analysis at a nanometric scale. The TEM analysis of study was carried out on TECHNAI G² F20 X-TWIN MAT 200Kv field emission transmission electron microscopy.

2.6. Proximate and ultimate analyses

Proximate and ultimate analyses were performed on coal samples based on ASTM Standards [12]. All runs were repeated to check the instrument's results repeatability and reproducibility.

3. Results and discussion

3.1. Mineralogy of coal

The XRD analytical results show that the pulverized coal used in the combustion process in the power station mainly composed of siliceous mineral such as quartz (SiO_2), kaolinite [$Al_2(SiO_2O_5)(OH)_4$)] and the non-siliceous mineral for instance potassium selenium chloride (K_2SeCl_6) and little quantities of pyrite (FeS_2) (Fig. 1). The mineral suites in the coal samples used in the present study are consistent with the previous studies [4, 13, 14, 15, 16, 17]. Kaolinite is uniformly distributed in the coal samples. This mineral is commonly present in coal as two species with different crystallinity, namely a low crystallinity detrital kaolinite and a high crystallinity neomorphic kaolinite. These genetic types have been described earlier [18, 19]. Kaolinite contains water bound within their lattices.

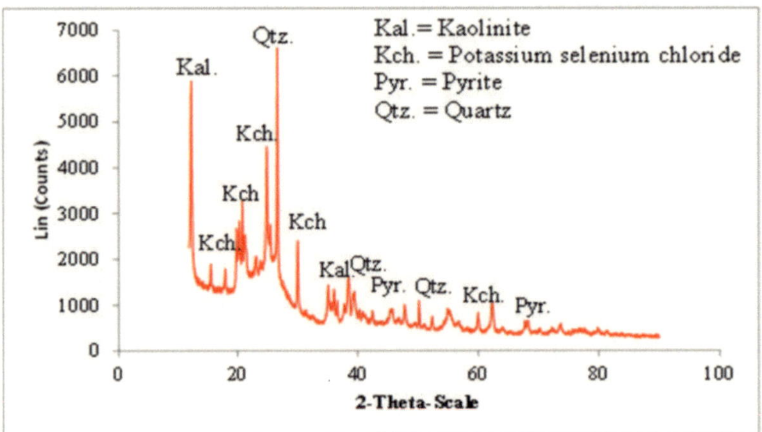

Figure 1. XRD spectra for the pulverised coal sample used in coal fired power station

All of the water is lost during the high-temperature ashing. Pyrite is the main species of sulfur oxidation in the coal samples studied. Pyrite occurs as typical syngenetic framboidal, euhedral and massive cell filling forms [20]. This mineral shows a highly inhomogeneous distribution in the coal samples. Pyrite is probably the most environmentally interesting mineral in the run of mine and beneficiation of coals and their generated wastes because of its propensity to oxidize during weathering and production of sulphuric acid.

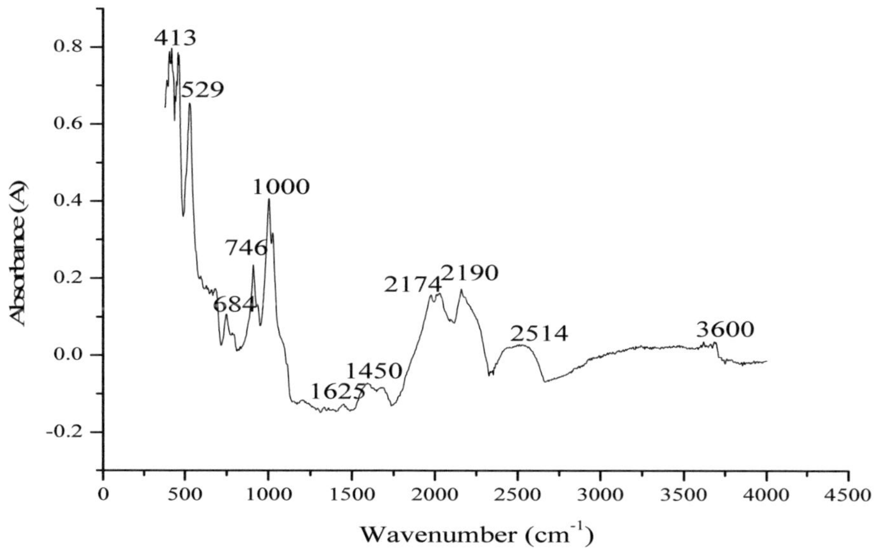

Figure 2. Infrared (FTIR) spectra of coal sample

Pyrite is transformed to hematite and sulfur dioxide during coal incineration at 815°C [20]. Some of the sulfur dioxide may remain combined with calcium in the ash, but much is lost [21]. The weathering of pyrite produces acid conditions that may leach trace elements associated with the pyrite and other constituents in the coal [22]. Quartz is the most common mineral in the coal samples studied. This mineral has mostly detrital genesis [23]. The shape of the quartz grains is rounded to semi-rounded, indicating an intensive transport before their deposition in the basin. The content of quartz is also high in the coal fly ash because this mineral is commonly stable/inert at combustion conditions (Fig. 3).

3.2. Mineralogy of weathered drilled ash cores

The mineralogical analysis by depth of the core ash samples was carried out with X- ray diffraction technique (XRD). This is to better understand the mineralogical changes (i.e. secondary phases) under the real dry disposal conditions. The experimental protocol for this section is presented in section 2.3. The results of the XRD analysis of samples of the drilled weathered dry disposed fly ash aged 2 week, 1 year and 20-year-old showed quartz (SiO_2) and mullite ($3Al_2O_3{\cdot}2SiO_2$) as main crystalline mineral phases (Fig. 3). Other minor mineral phases identified are; hematite (Fe_2O_3), calcite ($CaCO_3$), lime (CaO), anorthite ($CaAl_2Si_2O_8$), mica ($Ca(Mg,Al)_3(Al_3Si)O_{10}(OH)_2$) and enstatite ($Mg_2Si_2O_6$). This is in general agreement with mineralogy reported for other coal fly ashes [24, 25, 26, 27, 28, 29].The eight phases observed in every sample are considered to be characteristic phase assemblages for Class F fly ash. The XRD results obtained from 2-week-old dry disposed fly ash show similar mineralogical composition to the weathered ash except for the absence of calcite and mica. Lime (CaO) presence in coal fly ash may be due to the heat transformation of dolomite mineral or decarbonation of calcite entrained in feed stock coal [30]. The mullite present in fly ash was formed by the decomposition of kaolinite [31], which is entrapped in the parent coal. The gradual reduction in pore water pH is due to chemical interaction of fly ash with ingressed CO_2 and percolating rain water. Calcite formation is attributed to chemical weathering due to over time reduction in pore water pH. Previous study had proved that calcite precipitation in weathered fly ash is as a result of chemical interactions of calcium oxide (lime) rich fly ash with ingressed CO_2 [32]..

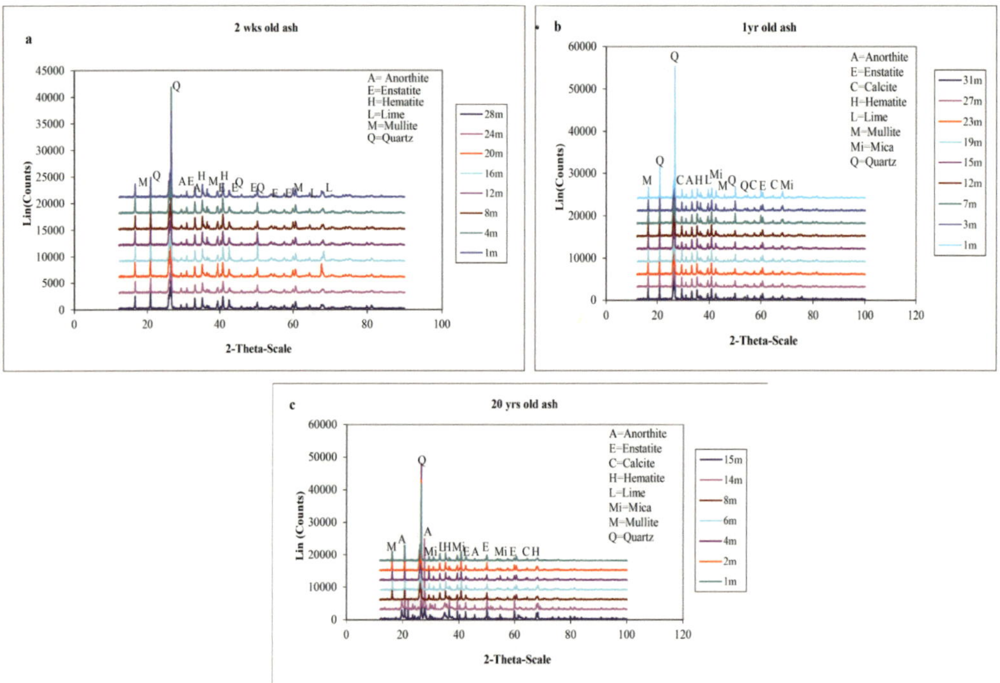

Figure 3. XRD spectra for the dry disposed ash dumps: (a) 2-week-old (T 87) not irrigated dry ash dump (b) 1-year-old (AMB 83) irrigated and quenched with high salient effluents (n = 2) (c) 20-year-old irrigated and quenched with fresh water.

Previous study has shown that the statistical variations in peak height on the same phase of ash samples could be used to assess the homogeneity and stability of different mineralogical phases [33]. A statistical consideration of the variations in the ashes peak heights for the different phases was used here to appraise the mineralogical distribution and chemical heterogeneity among the coal fly ash samples. Figures 4-6 presents the summary of the mean peak heights of mineral phases determined by XRD analysis of the 1 year, 8 year and 20-year-old ash core samples drawn from different depths of the dumps respectively. The eight phases observed in every sample are considered to be characteristic phase assemblages for Class F fly ash. This assemblage presented in Figure 4 consisted of quartz, anorthite (residual coal minerals), mullite, calcite, hematite, mica, enstatite and lime.

In the 1-year-old ash cores, the most prominent mineral phase is quartz having a peak of almost double that of the other mineral phases in all the core samples from all depths (0-31m). The peak height of quartz mineral showed a decreasing trend with increasing depth of 1-year-old ash dump (Figure 3). This observed trend showed strong correlation with flushing/leaching of SiO_2 and Al_2O_3 in the ash dump [34]. Thus, the quartz mineral peak height is indicative of rapid dissolution/weathering of aluminosilicate mineral within 1 year of ash dumping. Other mineral phases such ash mullite; calcite, hematite, enstatite, lime, anorthite and mica showed similar trend in the 1 year and 20-year-old ash dumps (Figures 4 and 6). The prominent presence of the quartz peak in the upper depths is due to flushing of other soluble matrix in the fly ash and the mineral phase quartz is easily detectable by XRD indicating there is relative increase in concentration in the upper depth

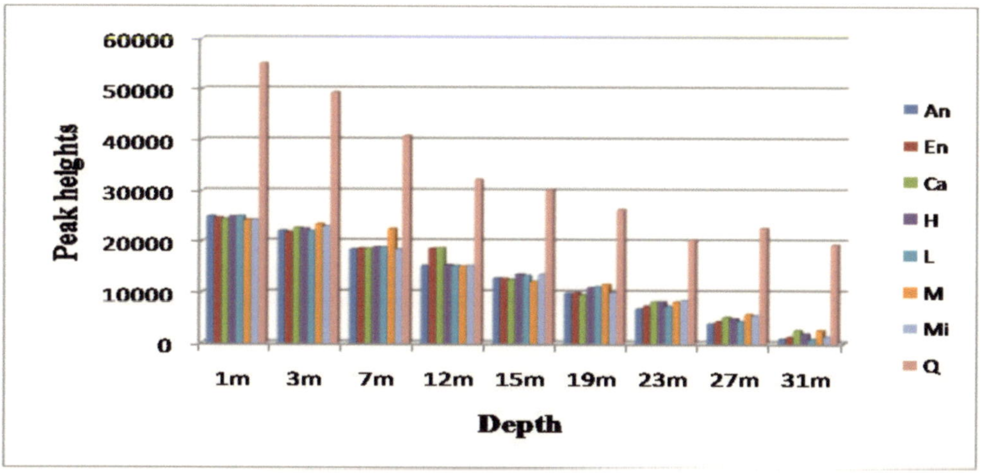

Figure 4. Bulk XRD mineral mean peak heights in 1-year-old (AMB 83) Tutuka ash cores (An=anorthite, En=enstatite, Ca=calcite, H=hematite, L=lime, M=mullite, Mi=mica, Q=quartz).

The mineral peak height in the 8-year-old section of the ash dump showed anomalous trend which may be ascribed to in-homogenous irrigation with high saline effluent (brine). This implied that the 8-year-old section has received much of high saline effluents than 1 year and 20-year-old sections of the ash dump.

The statistical result is shown in Table 1. The relative standard deviation (RSD) in the peak heights of 1-year-old drilled core (Figure 4 and Table 1) showed highly significant variations which could be classified into 2 groups:

$<$ 40 % Quartz

57-65 % Anorthite, Enstatite, Mica, Mullite, Hematite, Calcite and Lime

The relative standard deviation (RSD) of quartz phase (being $<$ 40 %) indicates less in-homogeneity among the coal fly ash mineral phases. There are however more variation (57-

65 %) in the peak heights of the following mineral phases, namely anorthite, enstatite, mica, mullite, calcite and lime.

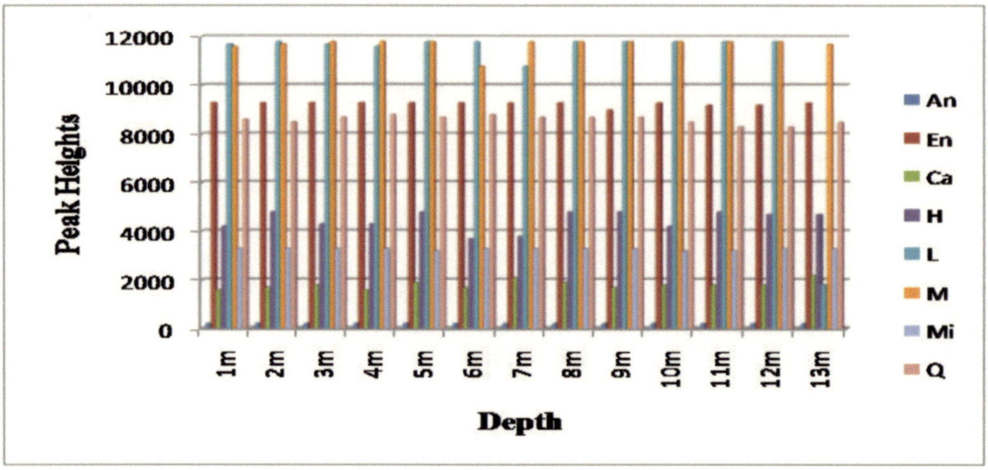

Figure 5. Bulk XRD mineral mean peak heights in 8-year-old (AMB 81) Tutuka ash cores (An=anorthite, En=enstatite, Ca=calcite, H=hematite, L=lime, M=mullite, Mi=mica, Q=quartz).

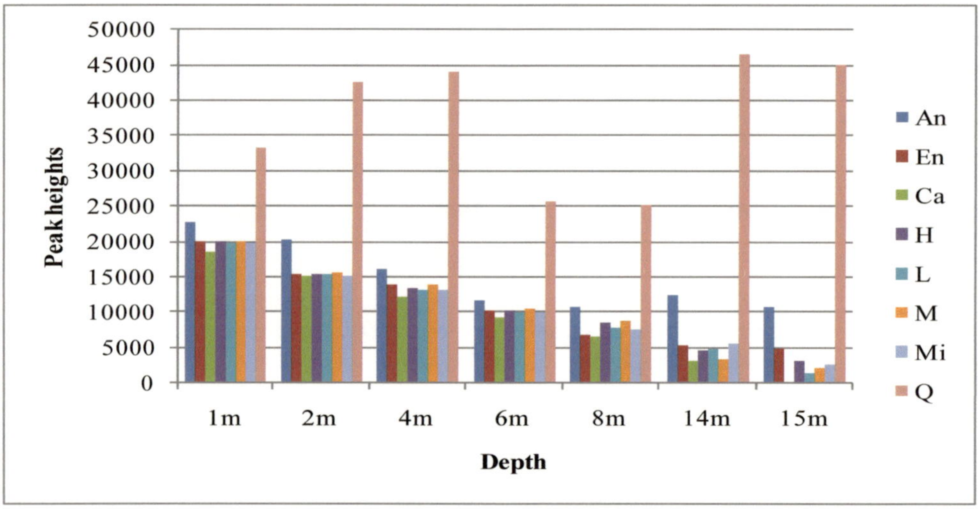

Figure 6. Bulk XRD mineral mean peak heights in 20-year-old (AMB 79) Tutuka ash cores (An=anorthite, En=enstatite, Ca=calcite, H=hematite, L=lime, M=mullite, Mi=mica, Q=quartz).

The in-homogeneous distribution of the calcite mineral phase could be attributed to continuous weathering occasioned by ingress of CO_2. There is a direct relationship between the calcite mineral phase peak heights and CaO (weight %). The depletion or enrichment of CaO (weight %) in coal fly ash agrees with the peak height of calcite which is an indication of the role of lime (CaO) in the formation of calcite ($CaCO_3$). The mica mineral phase also exhibited heterogeneous distribution in the drilled weathered cores which could be attributable to continuous weathering process.

1 year old core ash samples								
	An	En	Ca	H	L	M	Mi	Q
Mean	12776.3	13262.8	13614	13474.9	13128.2	13969	13310.4	33003.8
Stdev.	8198.9	8153.39	7813	7797.4	8057.18	7966.25	7759.5	12837.9
RSD %	64.17	61.48	57.39	57.87	61.37	57.03	58.30	38.90
8 year old ash core samples								
	An	En	Ca	H	L	M	Mi	Q
Mean	200.00	9261.54	1815.38	4453.85	10923.08	11692.31	3276.92	8600.00
Stdev.	0.00	83.56	170.28	381.53	2646.67	264.46	42.13	161.72
RSD %	0.00	0.90	9.38	8.57	24.23	2.26	1.29	1.88
20 year old ash core samples								
	An	En	Ca	H	L	M	Mi	Q
Mean	14955.4	10933.9	9302.0	10754.1	10291.0	10630.0	10513.0	37438.0
Stdev.	4862.3	5729.2	6515.9	5985.0	6284.1	6487.6	5980.7	9230.3
RSD %	32.5	52.4	70.0	55.7	61.1	61.0	56.9	24.7

[An=Anorthite, En=Enstatite, Q=Quartz, Mi=Mica, M=Mullite, Ca=Calcite, H=Hematite, L=Lime].

Table 1. Summary of the statistical analysis of the peak heights of various mineral phases in Tutuka fly ashes (1 year, 8 year and 20-year-old dry disposed cores)

Mullite showed variations (57-65 %) in the peak heights which might be due to the conversion of clays that contain < 60 % of Al_2O_3. The mullite peak heights depend on the amount of SiO_2 and Al_2O_3 in the mineral phase (Figure 4 and Table 1). Anorthite is a rare compositional variety of plagioclase feldspar (calcium rich end-member of plagioclase). Anorthite has been found in other fly ashes obtained from coals in which Ca is present in inherent minerals together with other minerals with which calcium reacts.

Anorthite showed variations (57-65 %) in peak height which could be due to chemical interactions of CaO, SiO_2 and Al_2O_3 in the mineral phase of coal fly ash samples. Enstatite is a magnesium silicate mineral. Enstatite shows variations (57-65 %) in peak heights due to prevailing chemical interactions of SiO_2 and MgO in the mineral phase of coal fly ash samples. Hematite (Fe_2O_3) is a heat transformation product of pyrite in feed coal and accordingly hematite was revealed by XRD in the ash core samples. Previous studies had shown that pyrite (FeS_2) was present in the feed coal as a fine-grained mineral [35]. Hematite (Fe_2O_3) phase showed variations (≈ 57 - 87 %) in peak heights distribution among the fly ash samples. This trend is an indication of the instability of the Fe containing mineral phases in the ash core samples of 1-year-old ash at the dry disposed fly ash dump site. The relative standard deviation (RSD) in the peak heights of the 8-year-old ash (Figure 5 and Table 1) showed low variations which could be classified into 3 groups:

< 3 % Anorthite, Enstatite, Quartz, Mica, Mulite,

< 10 % Calcite and Hematite

< 25 % Lime

The relative standard deviation (RSD) of anorthite, enstatite, quartz, mica and mullite (being < 3 %) indicates homogeneity among these coal fly ash mineral phases in this core. There was however more variation (< 25 %) in the mean peak heights of lime (CaO) in the 8-year-old ash samples. The in-homogeneous distribution of calcite and hematite mineral phase (< 10 %) could be attributed to continuous differential weathering occasioned by chemical interaction of fly ash with atmosphere, ingressed carbon dioxide and percolating rain water.

The relative standard deviation (RSD) in the peak heights of the 20-year-old ash (Figure 6 and Table 1) showed moderate variations which could be classified into 3 groups:

< 35 % Anorthite

< 60 % Enstatite, Mica and Hematite

≤ 70 % Lime, Mullite and Calcite

There is obvious similarity in the variation of mineral peak height of 20 year and 1-year-old drilled cores (Figures 4 & 6) which is attributed to in-homogeneity due to textural differences in 1 year and 20-year-old drilled ash cores [34]. Less variation in the mineral phase peak heights is observed with the age of the ashes due to dissolution/precipitation of secondary phases (stability of mineral phases) with time. The decrease or increase in the mean peak height of some minerals in the 3 drilled ash cores suggest variation in the steady state conditions at the interface of mineral particles due to reduction in the ash pore water pH [34]. The RSD in the mineral peak heights showed correlation with already significant change in chemistry of the 3 drilled ash core samples. This showed that significant flushing/leaching of major components of fly ash had taken place within 1 year of ash dumping due to continuous irrigation with high saline effluents [34].

Bulk chemical composition as determined by XRF analysis of all the coal fly ash samples also revealed major presence of MgO in fly ash which could result in the formation of enstatite ($Mg_2Si_2O_6$) upon concomitant depletion of quartz. Mullite and quartz were the species identified, quartz being the only original unaltered coal mineral phase present [36]. Mullite and hematite are products of the thermal transformation of some minerals present in the coal during combustion. Mullite ($3Al_2O_32SiO_2$) is a product of aluminosilicate transformation. It has been reported that mullite in fly ash is formed through the decomposition of kaolinite, $Al_2Si_2O_5(OH)_4$ [31, 35], which is entrapped in the parent coal. For UK fly ashes mullite is reported to form preferentially from kaolinite, whereas illite contributes towards the glass phase [35]. Calcite ($CaCO_3$) was found in all coal fly ash core samples. Calcite precipitation in weathered fly ash is as a result of chemical interactions of calcium oxide (lime) rich fly ash with ingressed CO_2 [32] during weathering. The high concentration of calcium oxide as evidenced by XRF analysis [34] of the core samples suggests possible secondary formation of anorthite ($CaAl_2Si_2O_8$), calcite, and lime in the fly ash dump [28]. Anorthite is a calcium mineral not present in coals. The formation of anorthite requires the mixing of separate calcium and aluminosilicate mineral domains. Thus, as calcium aluminosilicates are not found in fly ashes, their presence in coal-fired boiler deposits has been used to probe the mechanism of deposit formation. Studies of a range of coal fired boiler deposits, having different temperature histories, together with complementary investigations on mineral mixtures and coal ashes, have demonstrated that anorthite is formed via solid-state reactions and not by recrystallization from a homogeneous melt [38].

3.3. FTIR spectral analysis of the coal sample

FTIR was used to identify the mineral matter components removed and to monitor any changes in the functional groups resulting due to microbial treatment in the raw and bioleached coals. Figure 2 shows the FTIR spectra obtained for raw and bioleached coals. All peaks between 3600 cm^{-I} to 413 cm^{-1} were enlarged separately.

The high mineral content of the coals necessitated the analysis of the mineral matter more closely. Distinct peaks at the regions of 1000, 529 and 413 cm^{-1} are ascribed to kaolinite [39, 40]. In kaolinite (1:1 layer silicate), one of the silicate anions is replaced by a sheet of hydroxyl groups, and the layer units are linked by hydrogen bonds between the hydroxyl surface of the layer and the oxygen surface of the next layer. The high frequency OH^- vibrations occur at the region of 3600 cm^{-1}. In dioctahedral silicates, the degree of substitution of aluminium (A1) for silicon (Si) is lower than in trioctahedral silicates [41]. A perpendicular Si-O vibration causes absorption at 413 to 529 cm^{-1}. The Si-O-Si stretching vibrations give two bands at 1594 and 2174 cm^{-1} Si-O bending vibrations contribute to the strong absorption at 413 and 529 crn^{-1}.

The presence of quartz in the coal sample possibly gives rise to the IR spectrum with absorption frequency at 746 and 684 cm^{-1} [42, 43, 44]. The potassium selenium chlorides also absorb at 1625 and 1450 cm^{-1}. The small shoulder at 1450 cm^{-1} could be attributed to potassium selenium chlorides. The spectra indicated that the coal sample used in the present study had little iron sulphide. The iron sulphide (pyrite) is generally the most important of the iron-bearing minerals in coals (basic absorption frequency 413 crn^{-1}). Previous study established that the presence of quartz in the 2-week-old ash gives rise to IR spectrum with absorption frequency at 433-427, 770-764, and 991-996 cm^{-1}. Although, the presence of mullite is responsible for the series of bands around 540-532 and 1,413-1,000 cm^{-1} [45].

3.4. Microscopic study of coal and coal ash

SEM can provide size and morphology information of particles at submicron scale [46]. The size and morphological characteristics of coal and coal ash particles examined by SEM are exemplified in Fig. 7. SEM observations show that the coal and coal ash particles The coal ash (Fig. 7c, d) consists mostly of spherical shaped and aggregate that contains varying sizes and amount of particles. Conversely, the coal sample consisted of irregular shaped pyrite crystal coated with kaolinite (Fig 7a, b). TEM is the most powerful and appropriate technique for investigating the characteristics of nanoscale particles [46]. The morphology of coal and coal ash were identified using TEM. Transmission electron microscopy (TEM) is the most powerful and appropriate technique for investigating the characteristics of nanoscale particles [46]. The morphology of coal and coal ash were identified using TEM. The TEM images show that the coal and coal ash are made up of

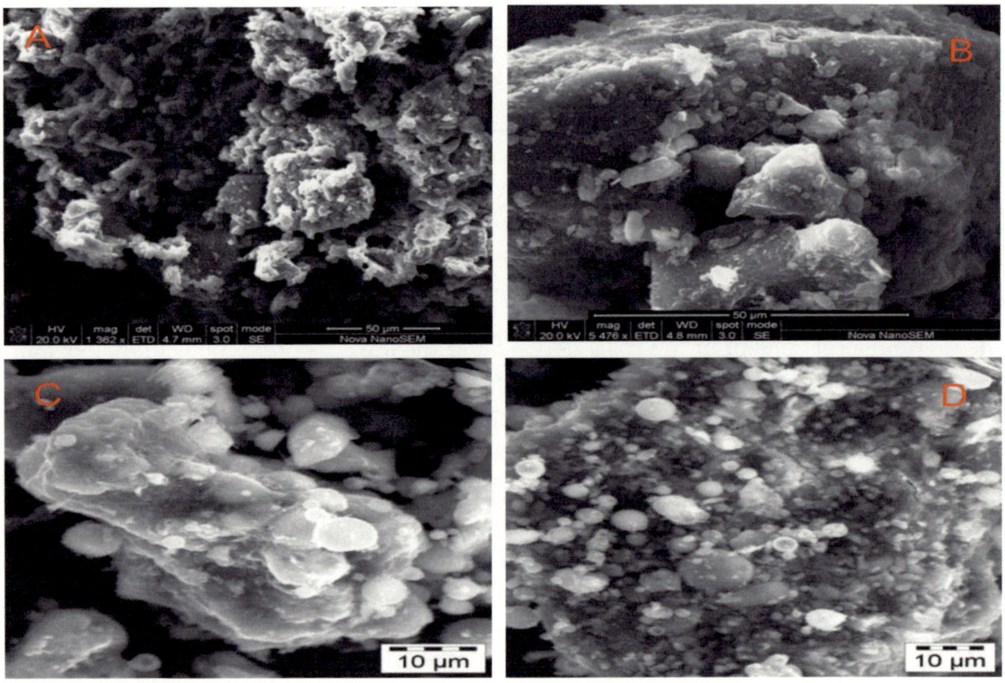

Figure 7. Scanning electron microscopy (SEM) of coal (a, b) and coal ash (c, d). Change the color of the letters so they are visible against the background of the photos

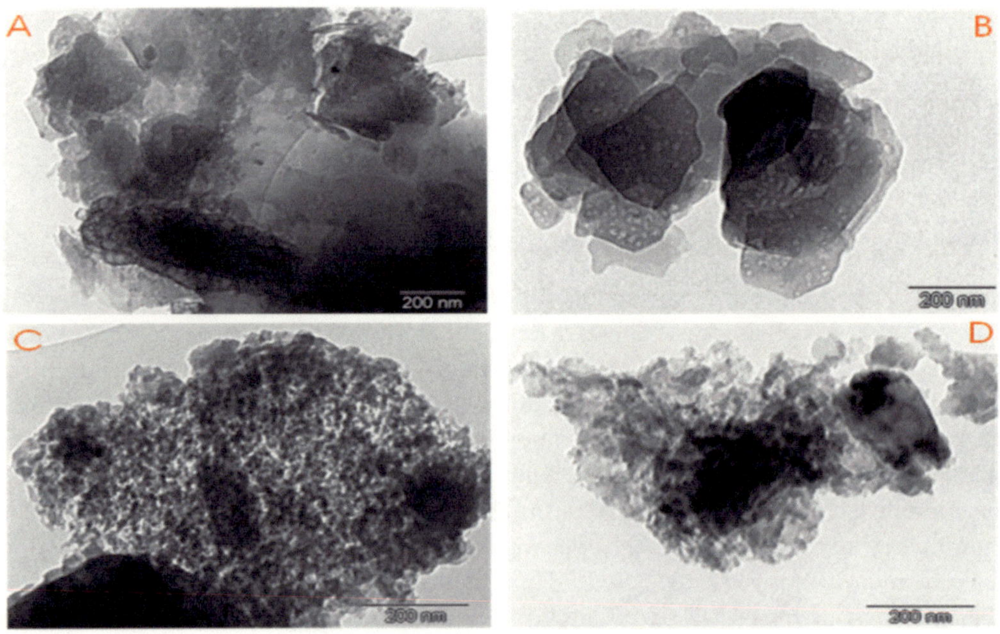

Figure 8. Transmission electron microscopy (TEM) of coal (a, b) and coal ash (c, d)

particles greater than 200μm. TEM images of coal ash sample showed nearly spherical shaped haematite structure (Fig. 8d), and cluster texture agglomeration of ultrafine particles (Fig. 8c). On the other hand, the TEM images shows that coal sample consist irregular shaped Fe-rich particles (i.e. pyrite) encrusted with Al, Si-rich particles (i.e. kaolinite) (Fig. 8 a, b). The metallifeorus particles such as Fe-rich particles, Al-rich particles and Si-rich particles are not uniformly distributed in the heterogeneous microstructure of coal ash [47, 48].

3.5. Geochemistry

Table 2 reports the chemical composition of coal and ash samples. The elements found in coals are commonly classified as major (> 1 wt. %), minor (1-0.1 wt. %) or trace (< 0.1 wt. %) elements. These elements may occur in both organic and inorganic constituents of coal and each element has dominant associations and affinities with different phases in coal [49]. The most abundant major components in both coal and ash samples are Si followed by Al, Fe and Ca. The least abundant components in the pulverised coal and fresh ash are Ti, Mg, Na, K, P, Cr and Mn. The bulk chemical composition and classification systems of coal fly ash always include data for LOI. The LOI and volatile (H_2O) components are relatively enriched in the pulverised coal sample. This combustion process thermally converts kaolinite to mullite as indicated in the Figures 1 & 2. This type of fly ash is principally composed of small (10 mm or less) glassy aluminosilicate spheres. The latter are formed by the rapid cooling of the molten mineral matter in the pulverized coal used in the power station boilers [50]. The ratio of Si/Al in the coal ash is ≥ 2 and thus can also be classified as silico-aluminate fly ash [51].

3.5.1. Major elements

Usually elements in coal occur either associated with the inorganic constituents (minerals) or with organic constituents [52]. The enrichment factor (EF) of the inorganic elements was calculated based on the method previously proposed by [53] (Table 2). Major elements such as Fe, Ca, Mg, Mn and Ti were as expected in coal samples but significantly enriched in the coal ash. Although, P, Na and K are slightly enriched in the coal ash samples used in this study. Enriched values in the ash were observed for environmentally significance trace elements.

3.5.2. Trace elements

The distribution of trace elements in coals used for electrical generation is of increasing importance in the assessment of environmental impacts from coal-fired power plants [54]. Trace elements such as U, Cr, Th, V, Ni, Cu, Zn, Rb, Sr, Mo and Sn are slightly enriched in the coal ash (Table 2).

The slight enrichment of these trace elements in the coal ash is attributed to the combustion process. Simultaneously, trace elements (such as Hf, Ta, Pb, Cr, Zr and Nb) showed significant enrichment in the coal ash. On the contrary, W, As, Cs and Ba are considerably enriched in the coal samples used in the present study. In summary, most of the determined trace elements were comparatively enriched in the coal ash when compared with the parent material (coal). Therefore, the trace elements relative enrichment in coal ash is attributed to the combustion process in the Tutuka power station.

Concentration (ppm)									
Element	LLD	Coal	Coal ash	EF	Element	LLD	Coal	Coal ash	EF
Si		79200.0	238700.0	0.94	As		231.08	47.67	0.064
Al		60400.0	126400.0	0.65	V	0.06	54.77	117.41	0.67
Fe		2600.0	37000.0	4.43	Cr	1.26	53.77	187.64	1.09
Ca		1300.0	41700.0	9.98	Co	0.02	24.25	17.53	0.22
Mg		600.0	5900.0	3.06	Ni	0.22	18.94	57.92	0.95
Mn		100.0	400.0	1.24	Cu	0.68	39.96	46.30	0.36
K		5100.0	6700.0	0.41	Zn	0.31	38.80	53.99	0.43
Na		500.0	600.0	1.06E-05	Rb	0.04	30.28	37.21	0.38
P		1200.00	2000.0	0.52	Sr	0.00	614.01	1270.70	0.644
Ti		4200.00	13500.0	1.00	Zr	0.01	99.09	392.21	1.23
U	0.002	3.82	10.25	0.84	Nb	0.01	10.35	35.45	1.07
Hf	0.01	2.84	10.77	1.18	Mo	0.00	3.24	6.31	0.12
Ta	0.004	0.82	2.65	1.01	Sn	0.07	4.07	8.95	0.68
W	0.00	119.17	6.99	0.02	Cs	0.01	6.16	5.92	0.30
Th	0.003	14.40	36.39	0.79	Ba	0.06	1200.18	1062.15	0.28
Pb	0.02	8.49	38.50	1.41					

* EF = $[(X) / (Ti)_{Ash} / (X) / (Ti)_{Coal}]$, where X means element (Ogugbuaja and James, 1995)

Table 2. Major and trace elements in the coal sample and coal ash from Tutuka Power Station ($n = 3$) (*LLD = Low level detection*)

3.5.3. Rare Earth Elements (REE)

Rare earth elements (REEs) contents in the coal used in the present study are summarized in Table 3. It shows that the bulk of REEs are found in high levels in coal ash when compared with the typical concentration in coal. Rare earth elements (REEs) such as La, Ce, Pr, Nd, Sm, Eu and Gd are slightly enriched in the coal ash. On the contrary, Lu, Y, Dy, Tb, Yb, Tm, Er and Ho are considerably enriched in the coal ash used in the present study. Enrichment of REEs in the coal ash disagreed with the previously held views [55, 56]. Consequently, the obvious enrichment of REEs in the coal ash used in the present study is attributed to the combustion conditions. Rare earth elements in coal appear to consist of a primary fraction which is associated with syngenetic mineral matter [57]. Another portion of the REE can be externally derived or mobilized when primary mineral matter is destroyed or modified.

3.6. Genetic features relations of coal and coal ash based on chemical composition

Minerals in coal are both detrital and authigenic in nature and their distribution in the inorganic matter are variable. Authigenic minerals in coal are mainly sulfides, carbonates and sulfates of Fe, Mg and Ca [58]. The chemical composition in this detrital authigenic index (DAI) also symbolizes the different index mineral (IM) in coal. For example, the oxides of Si, Al, K^+, Na^+, and Ti represent minerals and phases such as quartz, feldspars, clay and mica minerals (excluding some kaolinite and illite), volcanic glass, Al oxyhydroxide, and rutile-anatase-brookite, which commonly have dominant detrital genesis in coal. On the other hand, the oxides of Fe, Ca, Mg, S, P, and Mn represent minerals such as Fe-Mn sulphides; Ca-Fe-Mg sulphates, Ca-Mg-Fe-Mn carbonates and Ca-Fe phosphates, which commonly have dominant origin in coal [59]. Based on the ratio of detrital and authigenic minerals (DAI) some genetic information for the formation of fly ash could be deduced [51].

Concentration (ppm)				
Element	LLD	Coal	Coal ash	EF
La	0.002	39.90	91.36	0.71
Ce	0.004	91.61	182.42	0.62
Pr	0.002	9.46	19.72	0.65
Nd	0.016	30.84	71.76	0.72
Sm	0.009	5.30	14.38	0.84
Ho	0.003	0.67	2.35	1.09
Er	0.004	1.89	6.65	1.09
Tm	0.002	0.27	0.95	1.08
Yb	0.000	1.81	6.5	1.11
Eu	0.004	0.89	2.68	0.94
Gd	0.012	4.15	12.62	0.95
Tb	0.002	0.57	1.91	1.04
Dy	0.007	3.31	11.91	1.12
Y	0.01	17.50	64.87	1.15
Sc	0.12	9.66	26.50	0.85
Lu	0.002	0.25	0.93	1.15

* EF = $[(X) / (Ti)_{Ash} / (X) / (Ti)_{Coal}]$, where X means element (Ogugbuaja and James, 1995)

Table 3. Rare Earth Elements (REE) in the coal sample and coal ash from Tutuka Power Station (n =3) (*LLD = Low level detection*)

The main trend (Table 4) indicates that the coal used in the present study is a higher-ash coals which is enriched in elements associated with probable detrital minerals. Detrital minerals such as quartz, kaolinite, illite, acid plagioclases, muscovite, rutile, apatite and Fe and A1 oxyhydroxides are commonly stable minerals during coalification. Their proportions in coal may remain almost unchanged, while their total amount depends predominantly on the supply of clastic material into the peat swamp [23].

From Table 4, the proportion of detrital minerals is higher in coal sample used in the present study. It has been pointed out that the proportion of detrital minerals in coal increases [60]. The ratio of SiO_2/Al_2O_3 in the coal ash is ≥ 2 and thus can also be classified as silico-aluminate fly ash [51]. The bulk chemical composition and classification systems of coal fly ash always include data for LOI.

Major elements (%)										
Sample Name	SiO_2	Al_2O_3	Fe_2O_3	CaO	MgO	MnO	SiO_2/Al_2O_3	K_2O/Na_2O	$(MgO+CaO)/(K_2O+Na_2O)$	DAI
SAC	16.94	11.41	0.37	0.18	0.12	0.01	1.48	4.46	0.40	24.53
FA	51.05	23.88	5.29	5.84	1.26	0.05	2.14	5.37	7.39	5.97
Sample Name	Cr_2O_3	TiO_2	K_2O	Na_2O	P_2O_5	LOI	SO_3	Sum	CaO/MgO	
SAC	0.01	0.44	0.62	0.14	0.28	67.92	0.25	98.67	1.44	
FA	0.03	1.40	0.81	0.15	0.45	8.51	0.05	98.76	4.63	

DAI: $((SiO_2+Al_2O_3+K_2O+Na_2O+TiO_2) / (Fe_2O_3+CaO+MgO+SO_3+P_2O_5+MnO))$.

Table 4. Genetic features of coal and coal ash based on chemical composition

[61] classified fly ash based on the intersection of the sum of their major oxides: sialic: $SiO_2+Al_2O_3+TiO_2$; calcic: $CaO+MgO+NaO_2+K_2O$; and ferric: $Fe_2O_3+MnO+P_2O_5+SO_3$ in a ternary diagram. Based on the chemical composition of coal ash, about seven intermediate fly ash subgroups exists, such as sialic, ferrosialic, calsialic, ferrocalsialic, ferric, calcic and ferrocalcic [51] fly ash. The 1-year-old ash core samples were both sialic and ferrocalsialic in chemical composition (i.e. essentially Fe, Ca, Al and Si). Although, the 2 week and 20-year-old dry disposed ash core samples were sialic in chemical composition (i.e. essentially dominated by Al and Si) (Fig. 8). These trends show that in the 1-year-old drilled cores, there is already a significant change in chemistry of ash core due to rapid weathering or due to irrigation with high saline effluents. The coal fly ash transforms into a more clay-like material due to long-term mineralogical changes occasioned by the weathering process.

3.7. Proximate analysis and coal quality

The result obtained from proximate and ultimate analyses of pulverised coal sample is given in Table 5. The moisture and ash contents on dry and wet basis of pulverised coal sample (0.8 %; 94.43 %; 93.67 %) respectively. These values are higher than the Polish coals (0.58 %; 4.79 %) but the American coal (1.07 %; 5.77 %) was significantly higher in the moisture content. Some Nigerian coal deposits such as Lafia-Obi (2.91 %; 8.7 %) and Chikila coals (5.82 %; 14.9 %) also have considerably higher moisture content [62]. The relatively low moisture content in the pulverised coal sample represents a significant improvement in coal's quality because moisture affects the calorific value and the concentration of other constituents [63]. Nevertheless, the ash content of American coal, Lafia-Obi and Chikila coals are relatively lower than the ash content on dry basis of the pulverised coal used in this study. Similarly the low ash content is an improvement on the coking quality, low ash content is an essential requirement for coke making coals [63].

Therefore the pulverised coal used in this study may be expected not to have good coking qualities. An ash content of less than 10 % is recommended for a good coking coal (Bustin et al., 1985). Industrial experience indicates that a 1 wt. % increase of ash in the coke reduces metal production by 2 or 3 wt. % [65].

The volatile matter on dry and wet basis of pulverised coal sample is (5.6; 6.3) respectively (Table 5). The volatile matter of the pulverised coal sample used in this study is considerable lower than the American coal (31.36 %), Polish coal (32.61 %), Lafia-Obi (29.37 %) and Chikila (44.27 %) [66, 62]. Volatile matter, apart from its use in coal ranking, is one of the most important parameters used in determining their suitable applications [67]. Volatile matter does not form part of the coal; it is usually evolved as tar during carbonization. High-volatile bituminous coal due to its high volatile matter content generates high pressure during carbonization which is detrimental to the coke oven walls ([68, 69]. The above-mentioned data indicated that the coal used in the present study can be classified as medium volatile bituminous coal according to ASTM specification [70].

The elemental composition and elemental ratios of the coal sample used in this study are listed in Table 5. The obtained values for C, N and H contents are within the range observed for various types of coal [66, 62]. The C/N ratios of coal sample used in this study are higher than those reported for two south Brazilian coals [71]. On the other hand, similar values of H/C ratios observed in the present study have been obtained from two south Brazilian coals [71]. The fixed carbon of coal sample used in this study is 45.75 %. This is relatively higher than fixed carbons obtained from Chikila (40.83 %) coal. On the contrary, it is considerably lower than the fixed carbons in the Lafia-Obi (61.93 %), American (62.87 %) and Polish (62.60 %) coals [66, 62]. The carbon content of a coal is essential in coke making because it is the mass that forms the actual coke [72]. Therefore based on the fixed carbon the coal used in this study may be expected not to have good coking qualities.

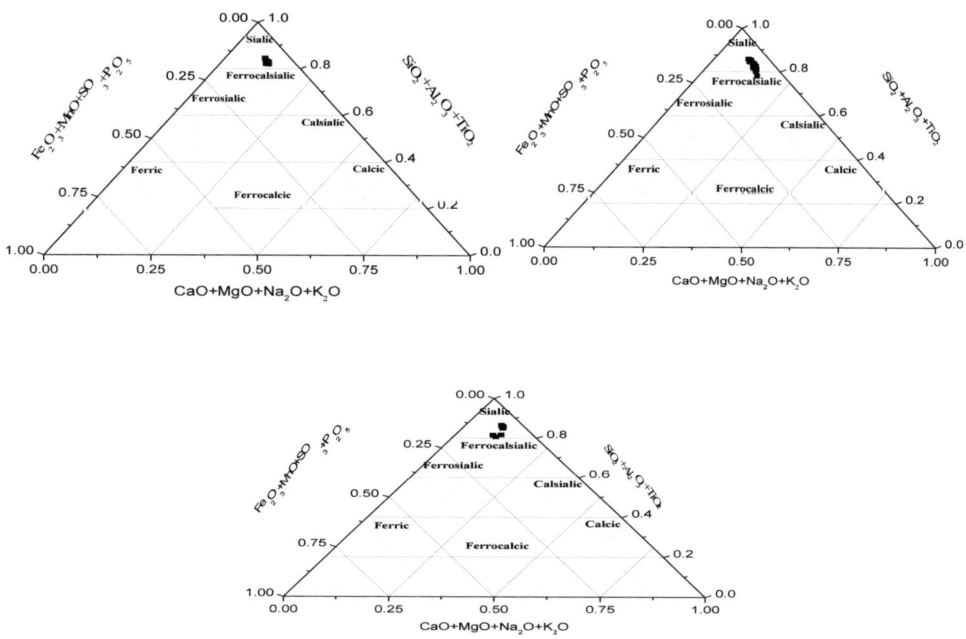

Figure 9. Ternary oxide plots for classification of the ash dumps: (a) 2-week-old (T 87) not irrigated dry ash dump (b) 1-year-old (AMB 83) irrigated and quenched with high salient effluents (n = 2) (c) 20-year-old irrigated and quenched with fresh water.

Proximate analysis of coal sample						
Sample	% Moisture	% Ash (dry basis)	% Ash (wet basis)	Volatile matter (dry)	Volatile matter (wet)	
SAC	0.8	94.43	93.67	5.6	6.3	
Ultimate analysis of coal sample						
Sample	N %	C %	H %	C/N	H/C	OM %
SAC	1.16	45.75	3.50	39.55	0.08	77.78

Note: OM % = C %*1.7.

Table 5. Proximate and ultimate analyses of South African coal sample *(n = 3)*

The organic matter content calculated (OM %) was calculated from the carbon content by multiplying with a value of 1.7 (Table 3). The derived organic matter content of the coal sample used in the present study is comparatively higher than Lafia-Obi coal (Jauro et al., 2008; Nasirudeen and Jauro, 2011).

4. Conclusions and summary

The XRD spectra showed that the coal sample mainly composed of siliceous mineral (such as quartz and kaolinite) and the non-siliceous mineral (such as potassium selenium chloride) and little quantities of pyrite. The results of the XRD analysis of samples of the drilled weathered dry disposed fly ash aged 2 week, 1 year and 20-year-old showed quartz and mullite as main crystalline mineral phases. Other minor mineral phases identified are; hematite, calcite, lime, anorthite, mica and enstatite.

The IR spectrum revealed the presence of quartz, kaolinite, potassium selenium chloride and pyrite in coal sample. SEM image of coal ash reveals spherical shaped and aggregate that contains varying sizes and quantity of particles. Conversely, the coal sample consists of irregular shaped pyrite crystal coated with kaolinite. TEM images of coal ash sample show nearly spherical shaped haematite structure and cluster texture agglomeration of ultrafine particles. Conversely, the TEM images of coal sample show irregular shaped Fe-rich particles (i.e. pyrite) encrusted with Al, Si-rich particles (i.e. kaolinite).

The main trend in the major oxides indicates that the coal used in the present study is a higher-ash coals which is enriched in elements associated with probable detrital minerals. The 1-year-old ash core samples were both sialic and ferrocalsialic in chemical composition (i.e. essentially Fe, Ca, Al and Si). Although, the 2 week and 20-year-old dry disposed ash core samples were sialic in chemical composition (i.e. essentially dominated by Al and Si).

Major elements such as Fe, Ca, Mg, Mn and Ti were as expected in coal samples but significantly enriched in the coal ash. Although, P, Na and K are slightly enriched in the coal ash samples used in this study.

Trace elements such as U, Cr, Th, V, Ni, Cu, Zn, Rb, Sr, Mo and Sn are slightly enriched in the coal ash. This slight enrichment of these trace elements in the coal ash is attributed to the combustion process. Nevertheless, trace elements (such as Hf, Ta, Pb, Cr, Zr and Nb) showed significant enrichment in the coal ash. On the contrary, W, As, Cs and Ba are considerably enriched in the coal samples used in the present study.

Rare earth elements (REEs) such as La, Ce, Pr, Nd, Sm, Eu and Gd are slightly enriched in the coal ash. On the contrary, Lu, Y, Dy, Tb, Yb, Tm, Er and Ho are considerably enriched in the coal ash used in the present study.

The proximate analysis revealed that the moisture content, ash content and volatile organic matter of pulverised coal used in this study is relatively low in values compare to the American coal, Polish coal, Lafia-Obi and Chikila coals. The ultimate analysis showed that the fixed carbon of coal sample used in this study is relatively higher than fixed carbons obtained from Chikila coal. On the contrary, it is comparatively lower than the fixed carbons in the Lafia-Obi, American and Polish coals

In conclusion, factors such as the nature of combustion process, type of coal and chemical interaction of fly ash with the ingressed CO_2 and percolating rain water would ultimately determine the mineralogy and chemical composition of coal combustion products.

Author details

S. A. Akinyemi, A. Akinlua
Fossil Fuel and Environmental Geochemistry Group, Department of Earth Sciences; University of the Western Cape, Bellville, South Africa

L. F. Petrik
Environmental and Nano Sciences Group, Department of Chemistry; University of the Western Cape, Bellville, South Africa

W. M. Gitari
Environmental Remediation and Water Pollution Chemistry Group, Department of Ecology and Resources Management, School of Environmental Studies, University of Venda. X5050, Thohoyandou, South Africa

5. References

[1] Jeffrey, L. S. Characterization of the coal resources of South Africa. The Journal of the South African Institute of Mining and Metallurgy 2005; 95-102.
[2] DME (Department of Minerals and Energy), Digest of South African Energy Statistics. Department of Minerals and Energy, Pretoria. 2005.
[3] Daniel, M. African coat supply prospects, lEA Coal Research, London, 1991.
[4] Snyman, C. P. and W. J. Botha, W. J. Coal in South Africa. Journal of African Earth Sciences, 1993; 16 171-180.
[5] Renton, J. J. Mineral matters in coal, In: Meyers, R. A. (Ed.). *Coal Structure*. New York Academy Press. 1982.

[6] Kolker, A.; Finkelman, R. B.; Palmer, C. A.; & Belkin, H. E. Arsenic, mercury and other trace metals in coal: Environmental and health implications (Abstract, International Ash Utilization Symposium, Lexington, KY); 2001.

[7] Burgers, C. L. Synthesis and characterization of sesquioxidic precipitates formed by the reaction of acid mine drainage with fly ash leachate. Unpublished M.Sc. Thesis, University of Stellenbosch, South Africa; 2002.

[8] Bezuidenhout, N. Chemical and mineralogical changes associated with leachate production at Kriel power station ash dam. Unpublished M.Sc. Thesis, University of Cape Town, Cape Town, South Africa; 1995.

[9] Petrik, L. F., White, R. A., Klink, M. J., Somerset, S. V., Burgers, L. C., and Fey, V. M.. Utilization of South African fly ash to treat acid coal mine drainage, and production of high quality zeolites from the residual solids. International Ash Utilization Symposium, Lexington, Kentucky; October 20–22. 2003.

[10] Eskom abridged annual report. (2009). Available at www.eskom.co.za, Accessed on 23rd May, 2010.

[11] Basham, K., Clark, M., France, T., and Harrison, P. What is fly ash? Fly ash is a by-product from burning pulverized coal in electric power generating plants. Concrete Construction Magazine; November 15. 2007.

[12] American Society for Testing and Materials. Annual Book of ASTM Standards, Section 5: Petroleum Products, Lubricants and Fossil Fuels. 5.05: Gaseous Fuels; Coal and Coke; 1992.\

[13] Vassilev, S. V. Trace elements in solid waste products from coal burning at some Bulgarian thermoelectric power stations. Fuel 1994a; 73 367-374.

[14] Querol, X., Fernández-Turiel, J. L., Lopez-Soler, A. Trace elements in coal and their behaviour during combustion in a large power station. Fuel 1995; 74 331.

[15] Vassileva, G. G., Vassilev, S. V. Behaviour of inorganic matter during heating of Bulgarian coals 1. Lignites. Fuel Processing Technology 2005; 86 1297-1333.

[16] Vassileva, C. G., Vassilev, S. V. Behaviour of inorganic matter during heating of Bulgarian coals 2. Sub bituminous and bituminous coals Fuel Processing Technology 2006; 87 1095–1116.

[17] Silva, L. F.O., Ward, C. R., Hower, J. C., Izquierdo, M., Waanders, F., Oliviera, M. L. S., Li, Z., Hatch, R. S., Querol, X. Mineralogy and Leaching Characteristics of Coal Ash a Major Brazilian Power Plant. Coal Combustion and Gasification Products 2010; 2 51-65.

[18] Ward, C. R., Spears, D. A., Booth, C. A., Staton, I. Mineral matter and trace elements in coals of the Gunnedah Basin, New South Wales, Australia. Int J Coal Geol. 1999; 40 (4) 281–308.

[19] Querol X, Alastuey A, Lopez-Soler A, Plana F, Zeng RS, Zhao J, et al. Geological controls on the quality of coals from the West Shandong mining district, Eastern China. Int J Coal Geol. 1999; 42 (1) 63–88.

[20] Liu, G., Vassilev, S. V., Gao, L., Zheng, L., Peng, Z. Mineral and chemical composition and some trace element contents in coals and coal ashes from Huaibei coal field, China. Energy Conversion and Management 2005; 46 2001–2009.

[21] Gluskoter, H. J. Mineral Matter and Trace Elements in Coal. In Trace Elements in Fuel; Babu, S.; Advances in Chemistry; American Chemical Society: Washington, DC. 1975.

[22] Liu, G. J., Yang, P. Y., Chou, C. L., Peng, Z. C. Petrographical and geochemical contrasts and environmentally significant trace elements in marine-influenced coal seams, Yanzhou Mining Area, China. J Asian Earth Sci. 2004; 24 (3) 491–506.

[23] Vassilev, S. V., Kitano, K., and Vassileva, C. G. Relations between ash yield and chemical and mineral composition of coals. Fuel 1997b; 76, (1) 3-8.

[24] McCarthy, G. J., Swanson, K. D., Keller, L. P., and Blatter, W. C.. Mineralogy of Western fly ash Cement and Concrete Research 1984; 14 471-478.

[25] Filippidis A., Georgakopoulos, A. Mineralogical and chemical investigation of fly ash from the Main and Northern lignite fields in Ptolemais, Greece. Fuel 1992; 71 373-376.

[26] Vassilev, S. V., Vassileva, C. G. Mineralogy of combustion wastes from coal-fired power stations. Fuel Process Technol. 1996a; 47 261–280.

[27] Sakorafa, V., Burragato, F., Michailidis, K. Mineralogy, geochemistry and physical properties of fly ash from the Megalopolis lignite fields, Peloponnese, Southern Greece. Fuel 1996; 75 419–23.

[28] Bayat, O. Characterization of Turkish fly ashes. Fuel 1998; 77 1059-1066.

[29] Koukouzas, N. K., Zeng, R., Perdikatsis, V., Xu, W., Emmanuel K. Kakaras, E. K. Mineralogy and geochemistry of Greek and Chinese coal fly ash. Fuel 2006; .85 2301–2309.

[30] Navarro, C. R., Agudo, E. R., Luque, A., Navarro, A. B. R., Huertas, M. O. "Thermal decomposition of calcite: Mechanisms of formation and textural evolution of CaO nanocrystals",American Mineralogist 2009; 94 578-593.

[31] White, S. C., and Case, E. D. Characterization of fly ash from coal-fired power plants. J. Mater. Sci. 1990; 25 5215–5219.

[32] Soong, Y., Fauth, D. L., Howard, B. H., Jones, J. R., Harrison, D. K., Goodman, A. L., Gray, M. L., and Frommell, E. A. CO_2 sequestration with brine solution and fly ashes. Energy Convers. Manage. 2006; 47 1676–1685.

[33] Stevenson, R. J., McCarthy, G. J. Mineralogy of Fixed-Bed Gasification Ash Derived from North Dakota Lignite. Mat. Res. Soc. Symp. Proc. 1985; Vol. 65.

[34] Akinyemi, S. A. Geochemical and mineralogical evaluation of toxic contaminants mobility in weathered coal fly ash: as a case study, Tutuka dump site, South Africa. Unpublished PhD thesis, University of the Western Cape, South Africa; 2011a.

[35] Spears, D. A. and Martinez-Tarazona, M. R. Geochemical and mineralogical characteristics of a power station feed-coal; Eggborough, England, Int. J. Coal Geol. 1993; 22 1-20.

[36] Bandopadhyay, A. K. "A study on the abundance of quartz in thermal coals of India and its relation to abrasion index: Development of predictive model for abrasion", International Journal of Coal Geology 2010; 84 63-69.

[37] Spears, D. A. Role of clay minerals in UK coal combustion. Applied Clay Science 2000; 16 87–95.

[38] Unsworth, J. F., Barratt, D. J., Park, D., Titchener, K. J. Ash formation during pulverized coal combustion. 2. The significance of crystalline anorthite in boiler deposits. Fuel 1988; 67 632- 642.
[39] Farmer, V. C. The infrared spectra of minerals. London: Mineralogical Society; 1987.
[40] Sharma, D.K. & Gihar, S. Chemical cleaning of low grade coals through alkali-acid leaching employing mild conditions under ambient pressure conditions. Fuel, 1991; 70 663-665.
[41] Sharrna, D. K and Wadhwa, G. Demineralization of coal by stepwise bioleaching: a comparative study of three Indian coals by Fourier Transform Infra-Red and X-ray diffraction techniques. World Journal of Microbiology & Biotechnology 1997; 13 29-36.
[42] Criado, M., Fernández-Jimènez, A., and Palomo, A. Alkali activation of fly ash: Effect of the SiO_2/Na_2O ratio. Part I: FTIR study. Microporous & Mesoporous Mater. 2007; 106 180–191.
[43] Fernandez-Carrasco, L., and Vazquez, E. Reactions of fly ash with calcium aluminate cement and calcium sulphate. Fuel 2009; 88 1533–1538.
[44] Mollah, M. Y. A., Promreuk, S., Schennach, R., Cocke, D. L., and Guler, R. Cristobalite formation from thermal treatment of Texas lignite fly ash. Fuel 1999; 78 1277–1282.
[45] Akinyemi, S. A., Akinlua, A., Gitari, W. M., and Petrik, L. F. Mineralogy and Mobility Patterns of Chemical Species in Weathered Coal Fly Ash Energy Sources, Part A 2011b; 33 768–784.
[46] Utsunomiya, S., Ewing, R. C. Application of high-angle annular dark field scanning transmission electron microscopy, scanning transmission electron microscopy energy-dispersive X-ray spectrometry, and energy-filtered transmission electron microscopy to the characterization of nanoparticles in the environment. Environmental Science & Technology 2003; 37 786–791.
[47] Silva L. F. O., Moreno, T., Querol, X. An introductory TEM study of Fe-nanominerals within coal fly ash. Science of the Total Environment 2009; 407 4972–4974.
[48] Akinyemi, S. A., Akinlua, A., W. M. Gitari, W. M., Nyale, S. M., Akinyeye, R. O., Petrik, L. F. An Investigative Study on the Chemical, Morphological and Mineralogical Alterations of Dry Disposed Fly Ash During Sequential Chemical Extraction. Energy Science and Technology 2012; 3 (1) 28-37.
[49] Vassilev, S. V., Vassileva, C. G. Geochemistry of coals, coal ashes and combustion wastes from coal-fired power stations. Fuel Processing Technology 1997a; 5 1 19-45.
[50] Foner. H. A., and Robl, 1 T. L. Coal Use and Fly Ash Disposal in Israel. Energeia 1997; 8, (5) 1-6.
[51] Vassilev S. V. and Vassileva, C. G. A new approach for the classification of coal fly ashes based on their origin, composition, properties and behaviour. Fuel 2007; 86 1490-1512.
[52] Zhang, J., Ren, D., Zheng, C., Zeng, R., Chou, C.-L., Liu, J. Trace element abundances of major minerals of Late Permian coal from southwestern Guizhou province, China. Int. J. Coal Geol. 2002; 53 55–64.

[53] Ogugbuaja, V. O., James, W. D. INAA multielemental analysis of Nigerian Bituminous Coal and Coal ash. Journal of Radioanalytical and Nuclear Chemistry, Articles 1995; 191 181-187.

[54] Karayigit, A.I., Gayer, R.A., Ortac, F.E., Goldsmith, S. Trace elements in the Lower Pliocene fossiliferous kangal lignites, Sivas, Turkey. Int. J. Coal Geol. 2001; .47 73–89.

[55] Swaine, D. J. Trace Elements in Coal. Butterworths, London. 1990; p. 278.

[56] Baioumy, H. M. Mineralogical and geochemical characterization of the Jurassic coal from Egypt. Journal of African Earth Sciences 2009; 54 75–84.

[57] Palmer, C. A., Lyons, P. C., Brown, Z.A., Mee, J. S. The use of trace element concentrations in vitrinite concentrates and companion whole coal (hvA bituminous) to determine organic and inorganic associations. GSA Spec. Pap. 1990; 248 55–62.

[58] Vassilev, S., Yossifova, M. and Vassileva, CInternational Journal of Coal Geology 1994b; 26 185.

[59] Vassilev S.V., Vassileva C. G. Occurrence, abundance and origin of minerals in coals and coal ashes. Fuel Proc. Technol. 1996b; 48 85-106.

[60] Finkelman, R. B. Scanning Microscopy 1988; 2 (1). 97.

[61] Roy, W. R., Griffin, R. A. A proposed classification system for coal fly ash in multidisciplinary research. Journal of Envir. Qual. 1982; 11 563-568.

[62] Nasirudeen, M. B. and Jauro, A. Quality of Some Nigerian Coals as Blending Stock in Metallurgical Coke Production. Journal of Minerals & Materials Characterization & Engineering 2011; 10, (1) 101-109.

[63] International Energy Agency (IEA)/Organization for Energy Co-operation and Development (OECD), Coal in the energy supply of India, Paris; 2002. p. 28.

[64] Bustin, R. M., Cameron, A. R., Greve, D. A. and Kalkreuth, W. D. Coal Petrology: Its Principles, methods and applications. Geological Association of Canada. Course Notes 1985; 3 230.

[65] Diez, M. A., Alvarez, R. and Barriocanal, C. "Coal for metallurgical coke production: Prediction of coke quality and future requirements for coke making." Int. J. Coal Geol. 2002; 50 289-412.

[66] Jauro, A., Chigozie , A. A. & Nasirudeen, M. B. Determination of selected metals in coal samples from Lafia-Obi and Chikila. Science World Journal 2008; 3 (2) 79-81.

[67] Peng Chen. "Petrographic Characteristics of Chinese coals and their applications in coal Utilization Processes." Fuel 2002; 81 11-12.

[68] Barriocanal, C., Patrick, J. W., and Walker, A. "The laboratory identification of dangerously coking coals." Fuel 1997; 77 881-884.

[69] Walker, R., Mastalerz, M., and Padgett, P. "Quality of selected coal seams from Indiana: Implications for carbonization." Int. J. of Coal Geol., 2001; 47 277-288.

[70] American Society for Testing and Materials. Annual Book of ASTM Standards, 5–6, p. 650. [71] Dick, D. P., Mangrich, A. S., Menezes, S. M. C. and Pereira, B. F. 2002. Chemical and Spectroscopical Characterization of Humic Acids from two South Brazilian Coals of Different Ranks J. Braz. Chem. Soc. 2002; 13 (2) 177-182.

[71] Price, J., Gransden, J., and Hampel, K. Microscopy, Chemistry and Rheology tools to determine Coal and Coke Characteristics. 1st McMaster's coke making course. McMaster's University Hamilton, Ontario, Canada; 1997.1-4 74.

Chapter 3

Kinetic Methods of Analysis with Potentiometric and Spectrophotometric Detectors – Our Laboratory Experiences

Njegomir Radić and Lea Kukoc-Modun

Additional information is available at the end of the chapter

http://dx.doi.org/10.5772/52185

1. Introduction

The basic types of reactions used for determinative purpose encompass the traditional four in equilibrium-based measurements: precipitation (ion exchange), acid-base (proton exchange), redox (electron exchange) and complexation (ligand exchange). These four basic types, or cases that can be reduced to them, are also found in kinetic-based measurements with some distinguishable trends. The influence of concentration on the position of a chemical equilibrium is described in quantitative terms by means of an equilibrium-constant expression. Such expressions are important because they permit the chemist to predict the direction and completeness of a chemical reaction. However, the size of one equilibrium constant tells us nothing about the rate (the kinetic) of the reaction. A large equilibrium constant does not imply that a reaction is fast. In fact, we sometimes encounter reactions that have highly favorable equilibrium constants but are of slight analytical use because their rates are low. Commonly used kinetic methods based on chemistry of reaction employed have been selected [1, 2].

Kinetic methods of analysis are based on the fact that for most reactions the rate of the reaction and the analytical signal increase with an increase of the analyte concentration. In kinetic methods, measurement of the analytical signal is made under dynamic conditions in which the concentrations of reactants and products are changing as a function of time.

Generally, in analytical chemistry many methods of analysis are based on the equilibrium state of the selected reaction. In contrast to kinetic methods, equilibrium or thermodynamic methods are performed on systems that have come to equilibrium or steady state, so that the analytical signal should be stable during measurements. Kinetic and equilibrium parts of the selected chemical reaction are illustrated in the figure 1.

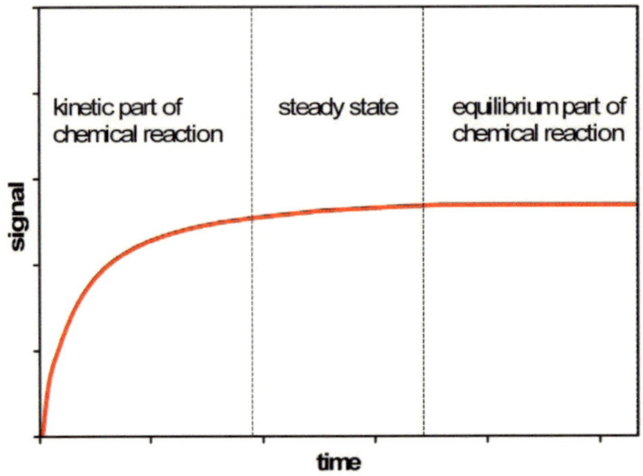

Figure 1. Kinetic, steady state and equilibrium parts of the selected chemical reaction

The most important advantage of kinetic method of the analysis is the ability to use chemical reaction that is slow to reach equilibrium. By using kinetic methods determination of a single species in a mixture may be possible when species have sufficient differences of reaction rates. In this chapter we present two analytical techniques where experimental measurements are made while analytical system is under kinetic control: i) chemical kinetic techniques, ii) flow injection analysis. The use of potentiometric and spectrophotometric detectors in kinetic methods are discussed. Also, the preparation and potential response of solid state potentiometric chemical sensors are described.

2. Kinetic methods of analysis with potentiometric detector

A potentiometric chemical sensor or ion-selective electrode confirms to the Nernst equation (1)

$$E = E^0 + k \log a_i \tag{1}$$

where E = the measured cell potential, E^0 = a constant for a given temperature, a_i = activity of an analyte ion in an aqueous solution and $k = RT \log(10)/nF$ where R is the gas constant, T is the absolute temperature, F is Faraday's constant and n is the number of electrons discharged or taken up by one ion (molecule) of an analyte. Usually, but not necessarily, n equals the charge (with sign) on the ionic form of the analyte. In practice, for constructing a calibration graph, it is normal to use solution concentrations instead of activities since concentration is more meaningful term to the analytical chemist than the activity. There are several points which should be noted from the response behaviors of the potentiometric chemical sensors when calibration graphs are constructed [3].

The electrode potential developed by an ion-selective electrode in a standardizing solution can vary by several millivolts per day for different measurements. For accurate measurement, therefore, the electrodes should be restandardized several times during the day. For a single determination, an error of 0.1 mV in measurements of the electrode potential results in an error of 0.39% in the value of monovalent anion activity [4]. Direct potentiometric measurements are usually time-consuming experiments.

Kinetic potentiometric methods are powerful tool for analysis, since they permit sensitive and selective determination of many samples within a few minutes with no sample pretreatment in many cases. The application of kinetic potentiometric methods offers some specific advantages over classical potentiometry, such as improved selectivity due to measurements of the evolution of the analytical signal with the reaction time. To construct calibration graphs the initial rate of the complex (product) formation reaction or change in potential during fixed time interval are used.

2.1. Use of potentiometric chemical sensors in aqueous solution

The fluoride ion-selective electrode (FISE) with LaF_3 membrane has proved to be one of the most successful ion-selective electrodes. FISE has a great ability to indirectly determine whole series of cations which form strong complexes with fluoride (such as Al^{3+}, Fe^{3+}, Ce^{4+}, Li^{+}, Th^{4+}, etc.). Combination of the simplicity of the kinetic method with the advantages of this sensor (low detection limit, high selectivity) produces an excellent analytical technique for determination of metal ions that form complexes with fluoride. Suitability of the FISE for monitoring a reasonable fast reaction of the formation of FeF^{2+} in acidic solution has been established by Srinivasan and Rechnitz [5]. Determination of Fe(III), based on monitoring of the formation of FeF^{2+} using FISE is described [6]. In this work, the kinetics of the FeF^{2+} formation reaction were studied in acidic solution (pH = 1.8; 2.5). The initial rates of iron(III)-fluoride complex formation in the solution, calculated from the non-steady-state potential values recorded after addition of Fe(III), were shown to be proportional to the analytical concentration of this ion in cell solution. The initial rate of the complex formation reaction, or change in potential during fixed time interval (1 minute), was used to construct calibration graphs. Good linearity (r = 0.9979) was achieved in the range of iron concentration from 3.5×10^{-5} to 1.4×10^{-5} mol L^{-1}. The described procedure can be usefully applied for the determination of free Fe(III) or labile Fe(III), as the fluoride may displace weaker ligands.

The determination of aluminium using FISE has mostly been performed in solutions buffered with acetate at pH 5, where fluorine is in the F^{-} form [7, 8]. Potential - time curves recorded during the Al-F complex formation reaction, using potentiometric cell with FISE, constitute the primary data in this study [7]. The initial rates decrease of the concentration of free fluoride ion were calculated and shown to be proportional to the amount of aluminium in reaction solution. The described method, based on the experimental observations, provides the determination of aluminium in the range from 8 to 300 nmol.

Aluminium(III) ions in aqueous solution show marked tendency to hydrolyse, with the formation of soluble, polynuclear hydoxo and aquo complexes and a precipitate of aluminium(III) hydroxide. The kinetic of the $AlF_i^{(3-i)+}$ formation reactions were also studied in acidic solution (pH 2) where both HF and HF_2^- exist, but parallel reactions of aluminium are avoided. The initial rates of aluminium-fluoride complex formation in this acidic solution, calculated from the non-steady-state potential values recorded after addition of aluminium, were shown to be proportional to the amount of this ion added [9].

The kinetic of aluminium fluoride complexation was studied in the large pH range. In the range of 0.9 – 1.5 [5], and from 2.9 to 4.9 [10].

Due to the toxicity of monomeric aluminium in free (aquo) and hydroxide forms, its rate of complexation with fluoride in the acidified aquatic environment is very important. A kinetic investigation of the rate and mechanism of reaction between Al(III) ions and fluoride in buffered aqueous solution (pH values 2 and 5) was described. The important paths of complex forming and the ecological importance of aluminium fluoride complexation in acidified aquatic environments were discussed [11]. In the laboratory solution, or in the aquatic environment, contains aluminium and fluoride ions, the following reactions may be considered to be the important path for aluminium-fluoride formation:

$$Al^{3+} + i\,F^- \rightleftarrows AlF_i^{(3-i)} \tag{2}$$

$$Al(OH)_j^{(3-j)} + i\,F^- + j\,H^+ \rightleftarrows AlF_i^{(3-i)} + j\,H_2O \tag{3}$$

$$Al^{3+} + i\,HF \rightleftarrows AlF_i^{(3-i)} + i\,H^+ \tag{4}$$

$$Al(OH)_j^{(3-j)} + i\,HF + j\,H^+ \rightleftarrows AlF_i^{(3-i)} + i\,H^+ + j\,H_2O \tag{5}$$

in these reactions, coordinated water has bee omitted for simplicity. Under the experimental conditions where $c_{Al} \gg c_F$, the formation of AlF^{2+} complex may be expected through one of the four possible paths (Eqs. 2–5), depending on the solution acidity. According to the theoretical consideration, after addition of aluminium, the recorded change in potential of the cell with FISE was higher at pH 2 than at pH 5. However, the rate of aluminium fluoride complexation is slightly slower at pH 2 than at pH 5.

Kinetic method of potentiometric determination of Fe(III) with a copper(II) selective electrode based on a metal displacement reaction is described [12]. Addition of various amounts of iron(III) to the buffered (pH 4) Cu(II)-EDTA cell solution alters the concentration of free copper(II) ion in the solution. EDTA is well known abbreviation for *ethylenediaminetetraacetic acid*, a compound that forms strong 1:1 complexes with most metal ions. EDTA is a hexaprotic system, designated H_6Y^{2+}. When iron(III) is added to a buffered aqueous solution containing CuY^{2-} species same cupric ion will be displaced because $K_{FeY^-} > K_{CuY^{2-}}$:

$$CuY^{2-} + Fe^{3+} \rightleftarrows FeY^- + Cu^{2+}$$

The above ligand exchange between two metals is often sluggish because the reaction involves breaking a series of coordinate bonds in succession [2]. As already noted [7], the rate of change in the potential, expressed as dE/dt, is directly proportional to the rate of change of the concentration of the potential determining ion, Cu^{2+} in this experiment, with time. The calculated values, $\Delta E/\Delta t$ *versus* log $c_{Fe(III)}$ was found to be linear for different concentrations of the Cu-EDTA complex, which was used as "kinetic substrate". The linear or analytical range for each tested concentrations of Cu-EDTA was close to one decade of iron concentration.

Kinetic potentiometric method for the determination of thiols (RSH): L-cysteine (cys), *N*-acetyl-L-cysteine (NAC), L-glutatione (glu) and D-penicillamine (pen) has been presented [13]. The proposed method is based on the reaction of formation the sparingly soluble salts, RSAg, between RSH and Ag^+. During the kinetic part of this reaction potential-time curves were recorded by using commercial iodide ion selective electrode with AgI-based sensitive membrane *versus* double-junction reference electrode as one potentiometric detector. The change of cell potential was continuously recorded at 3.0-sec interval. When the potential change, ΔE, recorded in 5th min. after RSH had been added in reaction solution, were plotted *versus* the negative logarithm of RSH concentration, p(RSH), rectilinear calibration graphs were obtained in the concentration ranges from 1.0×10^{-5} to 1.0×10^{-5} mol L^{-1}. The applicability of the proposed method was demonstrated by determination of chosen compounds in pharmaceutical dosage forms.

2.2. Use of potentiometric chemical sensors in non-aqueous solution

A change in solvent may cause changes in thermodynamic as well as kinetic properties of the selected chemical reaction. Also, the solubility of sensing membrane of one potentiometric chemical sensor, stability of the forming complexes, adsorption of reactants on the membrane and any undefined surface reaction may be strongly solvent dependent. Furthermore, the main properties of the used sensor which are important for analytical application such as sensitivity, selectivity response time and life-time, may be altered in non-aqueous solvents. In our experiments we have investigated different aqueous + organic solvent mixtures and their influence on thermodynamic and kinetic of the chemical reaction employed. Baumann and Wallace showed that cupric-selective electrode and a small amount of the copper(II)-EDTA complex could be used for the end-point detection in chelometric titrations of metals for which no electrode was available [14]. In the case of the compleximetric titration of mixtures of copper(II) an other metal ion in aqueous solution only the sum of both metals can be determined [15]. Titration measurements in the ethanol-aqueous media by using cupric ion-selective electrode as the titration sensor showed the possibility of direct determination of copper(II) in the presence of different quantity of iron(III) [16].

Sulfide ion-selective electrode was used as potentiometric sensor for determination of lead(II) in aqueous and nonaqueous medium. The initial rate of PbS formation was studied for series of solutions at various concentration of sodium sulfide and different pH values.

The measurements of lead sulfide formation in the presence of ethanol in 50% *V*/*V* were carried out in order to study the effect of organic solvent on the formation of the lead sulfide precipitate. After addition of Pb(II) ion, in water-ethanol mixtures, ethanol yielded higher potential jumps than in aqueous media [17].

Generally titrations in aqueous and nonaqueous media offer numerous advantages over direct potentiometry [18]. As it was mentioned, a change in solvent may cause changes in thermodynamic as well as in kinetic properties of the ions present. Also, the solubility of the FISE membrane, the stability of other metal fluorides, adsorption of fluoride ion and/or metal ions on the membrane and any undefined surface reaction, may be strongly solvent dependent. Furthermore, the main properties of the electrode used such as sensitivity, selectivity, response time and life time may be altered in non-aqueous solvents. Many papers have been concerned with the behavior of the FISE in a variety of organic solvents and their mixtures with water. Potentiometric titration of aluminium with fluoride in organic solvent + water mixtures by using electrochemical cell with FISE has been performed [19]. The potential break at the titration curve is not evident when titration is performed in aqueous solution. When the complexometric titration is performed in non-aqueous solution well defined S-shaped titration curves are obtained which suggest a simple stoichiometry of the titration reaction. In nonaqueous solutions, the formation of the complex with the maximum number of ligands (six) is presumably preferred. On the basis of potentiometric titration experiments the overall conditional formation constant of $AlF_i^{(3-i)}$ complexes have been calculated. Among the solvents tested (namely: ethanol, *p*-dioxane, methanol, n-propanol and tert-butanol) *p*-dioxane yielded a greater potential break than the other solvents and the measurements in mixtures with this solvent and ethanol also showed the best precision. The formation of aluminium hexafluoride complex in organic solvent + water mixtures may be accepted for the titration of higher concentration of aluminium (> 10^{-5} mol L^{-1}). However, at a low concentration of aluminium, the stoichiometric ratio between aluminium and fluoride was constant for a narrow range of aluminium concentrations and can be determined by experiment only.

The potentiometric determination of aluminium in 2-propanol + water mixtures was described [20]. The theoretical approach for the determination of aluminium using two potentiometric methods (potentiometric titration and analyte subtraction potentiometry) was discussed. The computed theoretical titration curves show that the equivalence point is signaled by great potential break only in media where aluminium forms hexafluoride complex. On the basis of the potentiometric titration and the known subtraction experiments in 2-propanol + water mixtures, the overall conditional constants $\{\beta'(AlF_6^{3-})\}$ were calculated. The calculated average β'-values are 10^{31} and 10^{33}, depending on the vol% of organic solvent, 50% and 67%. In the mixtures having a vol% of organic solvent of 50% or 67%, both methods can be applied for the determination of aluminium, *via* AlF_6^{3-} complex formation, in the concentration range from 1.0×10^{-4} to 1.0×10^{-3} mol L^{-1}.

3. Kinetic methods of analysis with spectrophotometric detector

In this chapter kinetic spectrophotometric methods are concerned to determination of thiols and similar compounds in pharmaceutical dosage forms. In fact, the spectrophotometric technique is the most widely used in pharmaceutical analysis, due to its inherent simplicity, economic advantage, and wide availability in most quality control laboratories. Kinetic spectrophotometric methods are becoming a great interest for the pharmaceutical analysis. The application of these methods offers some specific advantages over classical spectrophotometry, such as improved selectivity due to the measurement of the evolution of the absorbance with the reaction time. The literature is still poor regarding to analytical procedures based on kinetic spectrophotometry for the determination of drugs in pharmaceutical formulations. Surprisingly, to the authors' knowledge, there are only few published kinetic spectrophotometric methods for the determination of *N*-acetyl-L-cysteine (NAC) [21-23]. Also, only one of the cited methods for the determination of NAC has used Fe^{3+} and 2,4,6-trypyridyl-*s*-triazine (TPTZ) as a reagent solution. The reported method [23] is based on a coupled redox-complexation reaction. In the first (redox) step of the reaction, NAC (RSH compound) reduces Fe^{3+} to Fe^{2+} (Eq. (6)). In the second step of the reaction, the reduced Fe^{2+} is rapidly converted to the highly stable, deep-blue coloured $Fe(TPTZ)_2^{2+}$ complex (Eq. (7)) with λ_{max} at 593 nm:

$$2Fe^{3+} + 2RSH \rightleftarrows 2Fe^{2+} + RSSR + 2H^{+} \quad (6)$$

$$Fe^{2+} + 2TPTZ \rightleftarrows Fe(TPTZ)_2^{2+} \quad (7)$$

The initial rate and fixed-time (at 5 min) methods were utilized in this experiment. Both methods can be easily applied to the determination of NAC in pure form or in tablets. In addition, the proposed methods are sensitive enough to enable the determination of near nanomole amounts of the NAC without expensive instruments and/or critical analytical reagents. The kinetic manifold for a spectrophotometric determination of NAC or other thiols is shown in Fig. 2.

Kinetic spectrophotometric method for the determination of tiopronin {*N*-(2-mercaptopropionyl)-glycine, MPG} has been developed and validated. This method is also based on the coupled redox-complexation reaction (Eqs. 6, 7.) [24]. The use of TPTZ as chromogenic reagent has improved selectivity, linearity and sensitivity of measurements. The method was successfully applied for determination of MPG in pharmaceutical formulations.

The initial rate and fixed time (at 3 min) methods were utilized for constructing the calibration graphs. The graphs were linear in concentration ranges from 1.0×10^{-6} to 1.0×10^{-4} mol L^{-1} for both methods with limits of detection 1.3×10^{-7} mol L^{-1} and 7.5×10^{-8} mol L^{-1} for the initial rate and fixed time method, respectively. Under the optimum conditions, the absorbance-time curves for the reaction at varying MPG concentrations (1.0×10^{-6} to 1.0×10^{-4} mol L^{-1}) with the fixed concentration of Fe(III) (5.0×10^{-4} mol L^{-1}) and TPTZ (5.0×10^{-4} mol L^{-1}) were generated (Fig. 3).

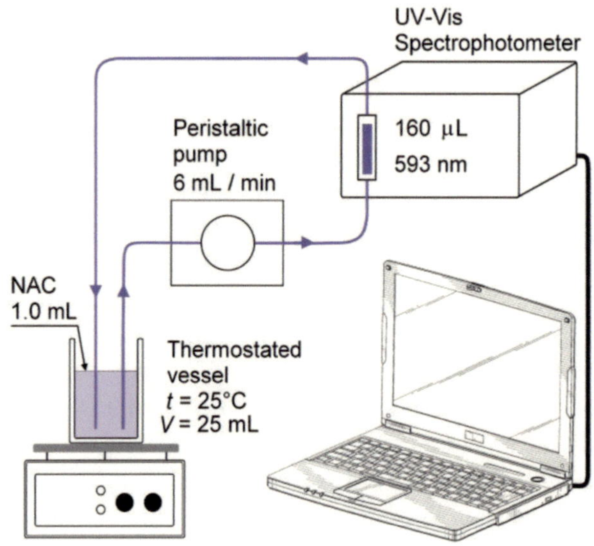

Figure 2. Kinetic manifold for the spectrophotometric determination of RSH. $\lambda = 593$ nm is for $Fe(TPTZ)_2^{2+}$ complex.

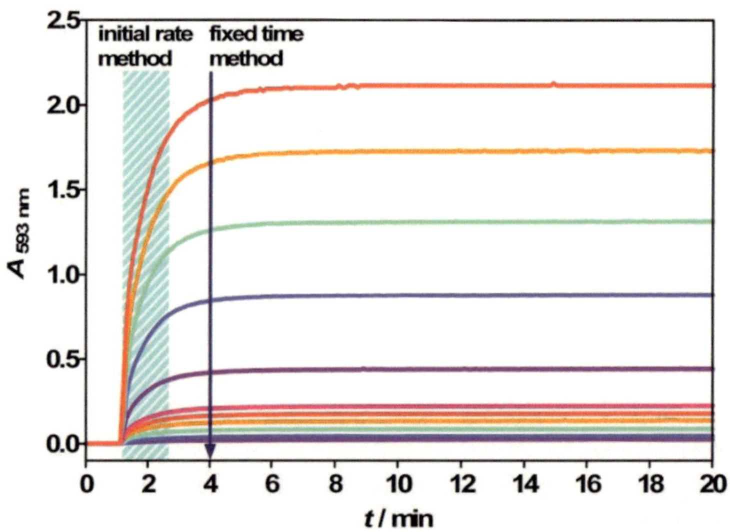

Figure 3. Absorbance as the function of time for the coupled redox-complexation reaction, measured at different MPG concentrations (1.0×10^{-6} to 1.0×10^{-4} mol L^{-1}) - Experimental conditions: $c(Fe^{3+}) = 5.0 \times 10^{-4}$ mol L^{-1}, $c(TPTZ) = 5.0 \times 10^{-4}$ mol L^{-1}, pH = 3.6, t = 25°C, analyte added 1 min after beginning of the measurement.

The initial reaction rates (K) were determined from the slopes of these curves. The logarithms of the reaction rates (log K) were plotted as a function of logarithms of MPG concentrations (log c) (Fig.4). The regression analysis for the values was performed by fitting the data to the following equation:

$$\log K = \log k' + n\log c \quad (8)$$

where K is reaction rate, k' is the rate constant, c is the molar concentration of MPG, and n (slope of the regression line) is the order of the reaction. A straight line with slope values of 0.9686 (≈ 1) was obtained confirming the first order reaction. However under the optimized reaction conditions the concentrations of Fe(III) and TPTZ were much higher than concentrations of MPG in the reaction solution. Therefore, the reaction was regarded as a pseudo-first order reaction.

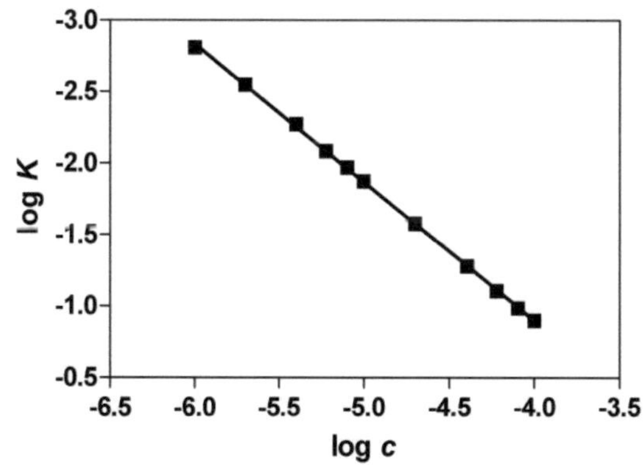

Figure 4. Linear plot for log c vs. log K for the kinetic reaction of MPG with Fe(III) (5.0×10^{-4} mol L^{-1}) and TPTZ (5.0×10^{-4} mol L^{-1}). c is [MPG]: (1.0×10^{-6} to 1.0×10^{-4} mol L^{-1}); K is the reaction rate (s^{-1}).

A simple kinetic method for the spectrophotometric determination of L-ascorbic acid (AA) and thiols (RSH) in pharmaceutical dosage forms, based on a redox reaction of these compounds with Fe(III) in the presence of 1,10-phenantroline (phen) at pH = 2.8 has been described [21].

Before RSH or AA is added to the reaction solution, Fe(III) and phen have formed a stable complex, $Fe(phen)_3^{3+}$.

The mechanism of redox-reaction of AA or RSH with the formed complex $Fe(phen)_3^{3+}$ may be written as:

$$H_2A + 2\,Fe(phen)_3^{3+} \rightleftarrows DA + 2\,Fe(phen)_3^{2+} + 2\,H^+ \quad (9)$$

$$2\,RSH + 2\,Fe(phen)_3^{3+} \rightleftarrows RSSR + 2\,Fe(phen)_3^{2+} + 2\,H^+ \quad (10)$$

where H_2A is the reduced form of AA and DA is dehydrogenized AA.

Catalytic-effect Cu(II) proposed by Teshima et al. [25] was applied to enhance analytical signal for slow redox-reaction. The catalytic effect of Cu^{2+} on redox-reaction thiols (RSH) with the Fe(III)-phen complex may be written as:

$$2\ \mathrm{RSH} + 2\ \mathrm{Cu}^{2+} \rightleftarrows \mathrm{RSSR} + 2\ \mathrm{Cu}^{+} + 2\ \mathrm{H}^{+} \tag{11}$$

$$2\ \mathrm{Cu}^{+} + 2\ \mathrm{Fe}\left(\mathrm{phen}\right)_3^{3+} \rightleftarrows 2\ \mathrm{Cu}^{2+} + 2\ \mathrm{Fe}\left(\mathrm{phen}\right)_3^{2+} \tag{12}$$

An orange-red iron(II)-phen complex produced by the reaction in Equation (12) absorbs at $\lambda = 510$ nm and Cu^{2+} ions will turn back to the reaction with RSH.

The rates and mechanisms of the chemical reactions, on which all these determinations are based, play a fundamental role in the development of an analytical (spectrophotometric) signal. Therefore, it was greatly important to establish the kinetics of chemical reactions applied for developing the proposed method. This investigation is also important for the optimization of the flow injection method used for the determination of the same compounds.

4. Flow injection analysis

In flow-injection analysis (FIA) the sample (analyte) is injected into a continuously flowing carrier stream, where mixing of sample and analyte with reagent(s) in the stream are controlled by the kinetic processes of dispersion and diffusion. Since the concept of FIA was first introduced in 1975 [26], it has had a profound impact on how modern analytical procedures are implemented. FIA with different detector is rapidly developing into a powerful analytical tool with many merits, such as broad scope and rapid sample throughput. The analytical signal monitored by a suitable detection device, is always a result of two *kinetic* processes that occur simultaneously, namely the *physical* process of zone dispersion and the superimposed *chemical* processes resulting from reaction between analyte and reagent species. As a result of growing environmental demands for reduced consumption of sample and reagent solutions, the first generation of FIA, which utilizes continuous pumping of carrier and reagent solutions, was supplemented in 1990 by the second generation, termed sequential injection analysis (SIA). In 2000, the third generation of FIA, the so-called lab-on-valve (LOV) was appeared [27].

4.1. FIA-methods with potentiometric detector

Over many years, there has been a great deal of research and development in FIA using ion-selective electrodes as detectors. A different design of flow-through potentiometric sensors has been investigated, but the incorporation of a tubular ion-selective electrode into the conduits of FIA has been used as a nearly ideal configuration, because the hydrodynamic flow conditions can be kept constant throughout the flow system [28].

For the determination of compounds containing sulphur a simple FIA system was developed. A simple tubular solid-state electrode with an AgI-based membrane hydrophobized by PTFE (powdered Teflon) was constructed and incorporated into a flow-injection system. The flow system and the configuration of the constructed tubular flow-through electrode unit have been described [29, 30].

For the experimental measurements, two-channel FIA setup has bee used. The tubular electrode and reference electrode are located downstream after mixing two channels. A constant representing a dilution of the sample and/or reagent after mixing of two solutions depends on the flow rates in channels and can be calculated as it has been shown [30].

In this experiment, the iodide electrode with (Ag_2S+AgI+PTFE)-membrane responds primarily to the activity of the silver ion at the sample solution–electrode membrane interface downstream after a confluence point of two channels. The preparation and performance of a silver iodide-based pellet hydrophobised by PTFE have been described [31].

In FIA experiment, using two-line flow manifold, the potential of the cell with the sensing tubular electrode is given by

$$E_1 = E^{'} + S \log a_{Ag^+} = E^{'} + S \log \left(c_{Ag^+} \cdot m \cdot \alpha_{Ag^+} \cdot f_{Ag^+} \right) \tag{13}$$

where $S, c, m, \alpha,$ and f denote the response slope of the electrode, the total or analytical concentration of silver ions in reagent solution, the dilution constant, the fraction of Ag^+, and the activity coefficient, respectively.

In the absence of ions in the streaming solution that form sparingly soluble silver salts or stable silver complexes and at constant ionic strength, the potential of the sensor can be expressed by the following equation:

$$E_1 = E^{''} + S \log \left(c_{Ag^{'}} \cdot m \right) \tag{14}$$

When a sample containing compound with sulfur (designated also as RSH) at a sufficiently high concentration to cause precipitation of RSAg is injected into the carrier stream, the silver ion concentration will be lowered to a new value. If $c_{RSH} \cdot d \cdot m \cdot \gg c_{Ag^+} \cdot m$, where dispersion of the sample is represented by the constant d, the free silver ion concentration at equilibrium can be analyzed and expressed as follows:

$$\left[Ag^+\right] = K_{sp,\, RSAg} / \left[RS^-\right] \tag{15}$$

$$c_{RSH} = \left[RS^-\right] + \left[RSH\right] = \left[RS^-\right] + \frac{\left[RS^-\right] \cdot \left[H^+\right]}{K_{a,\, RSH}} \tag{16}$$

$$\alpha_{RS^-} = \frac{\left[RS^-\right]}{c_{RSH}} = \frac{\left[RS^-\right]}{\left[RS^-\right] + \frac{\left[RS^-\right] \cdot \left[H^+\right]}{K_{a,\, RSH}}} = \frac{K_{a,\, RSH}}{K_{a,\, RSH} + \left[H^+\right]} \tag{17}$$

$$\left[RS^-\right] = c_{RSH} \cdot \alpha_{RS^-} \tag{18}$$

$$\left[Ag^{+}\right]=K_{sp,\ RSAg}\bigg/\left(c_{RSH}\cdot\frac{K_{a,\ RSH}}{K_{a,\ RSH}+\left[H^{+}\right]}\right) \tag{19}$$

$$\left[Ag^{+}\right]=K_{sp,\ RSAg}\cdot\frac{K_{a,\ RSH}+\left[H^{+}\right]}{K_{a,\ RSH}}\bigg/c_{RSH}\cdot d\cdot m \tag{20}$$

where $K_{sp,\ RSAg}$ is the solubility product of silver salt, while $K_{a,\ RSH}$ is the dissociation constant of sulfhydryl group, $K_{a,\ RSH}=\left[RS^{-}\right]\cdot\left[H^{+}\right]/\left[RSH\right]$.

In the flow-injection measurements of compounds with highly reactive sulfhydryl group, the potential of the peak may be described by the following equation:

$$E_{p}=E^{''}+S\log\left\{K_{sp,\ RSAg}\cdot\frac{K_{a,\ RSH}+\left[H^{+}\right]}{K_{a,\ RSH}}\bigg/c_{RSH}\cdot d\cdot m\right\} \tag{21}$$

The peak height h in these measurements is equal to the potential difference:

$$h=E_{1}-E_{p} \tag{22}$$

and using equations (14) and (21), one can obtain an equation for peak height. Hence, f, d, m, $\left[H^{+}\right]$, and $c_{Ag^{+}}$ are kept constant, and $c_{RSH}\cdot d\cdot m\gg c_{Ag^{+}}\cdot m$, a linear dependence between the peak height an logarithm of concentration of RSH with the slope of 59 mV $\left\{p\left(RSH\right)\right\}^{-1}$, can be obtained.

Application of a FIA system was exemplified by the determination of different compounds containing sulfur in 0.1 mol L^{-1} $HClO_4$ as a supporting electrolyte. For compounds with –SH group, a rectilinear calibration graph was obtained. The experimental slope was in good agreement with the theoretical value postulated on the precipitation process and formation of RSAg into the carrier stream or at the sensing part of the detector.

The equilibrium concentration of Ag^{+} ions will also be lowered if a sample contains RSH forms $Ag\left(SR\right)_{i}^{(1-i)}$ complexes instead of precipitation. Hence, if injected concentration of RSH is much higher than silver concentration in streaming solution, the potential of the peak may be described by the following equation:

$$E_{p}=E^{''}+S\log\left(c_{Ag^{+}}\cdot m\cdot\alpha_{Ag^{+}}\right) \tag{23}$$

$$\alpha_{Ag^{+}}=\frac{1}{1+\beta_{1}\left[RS^{-}\right]+\beta_{2}\left[RS^{-}\right]^{2}+\ \cdots\ +\beta_{i}\left[RS^{-}\right]^{i}} \tag{24}$$

where β is the stability constant and $[RS^{-}]$ is the free concentration of ligand. The concentration of ligand can be expressed by

$$\left[RS^-\right]=\left(c_{RSH}\cdot d\cdot m\cdot\frac{K_{a,\,RSH}}{K_{a,\,RSH}+\left[H^+\right]}\right) \tag{25}$$

If d, m, $\left[H^+\right]$ and c_{Ag^+} are kept constant and $c_{RSH}\cdot d\cdot m \gg c_{Ag^+}\cdot m$, a linear dependence between the peak height and logarithm of c_{RSH} may be obtained, but only if in the denominator of equation (24) one term predominates and "1" can be neglected. The slope of the potentiometric response will be iS mV $\{p(RSH)\}^{-1}$, where i is the number of ligands in the predominant complex. As it has been discussed [30], the solubility product of RSAg or the stability constant of $Ag(SR)_i^{(1-i)}$ complex can be calculated when a continuous-flow instead of a flow-injection technique was applied.

Potentiometric determination of penicillamine (pen, RSH) was described based on a batch experiment and FIA method [32]. Also, the solubility product $K_{sp,\,RSAg}$ was determined using experimental values recorded both by batch measurement and by the continuous-flow experiment. The mean value obtained by different measurements and using a membrane of the same composition (AgI+Ag_2S+PTFE) was $K_{sp,\,RSAg}=(1.4\pm0.1)\times10^{-20}$.

The preparation of a tubular electrode has been extended by means of chemical pretreatment of a silver tube with mercuric chloride and iodide solution. In this treatment AgI-sensing layer on the inner surface of tube was formed [33]. The electrode was used as a potentiometric sensor for the determination of ascorbic acid (vitamin C), glutathione and cysteine in batch and FIA experiments.

FIA system with cascade flow cell equipped with commercial FISE as detector has been described [34]. This system was applied for the determination of iron in the range of concentration from 1.0×10^{-4} to 1.0×10^{-1} mol L^{-1}.

4.2. FIA-methods with spectrophotometric detector

Recently, more strict regulation related to the quality control in pharmaceuticals has led to an increase of demands on automation of the analytical assays carried out in appropriate control laboratories. The FIA became a versatile instrumental tool that contributed substantially to the development of automation in pharmaceutical analysis due to its simplicity, low cost and relatively short analysis time. A simple, rapid and sensitive flow-injection spectrophotometric method for the determination of NAC has been successfully developed and validated [35]. In this work TPTZ was proposed as a chromogenic reagent for the determination of NAC in aqueous laboratory samples, instead of frequently employed 1,10-phenantroline. Reaction mechanism of the method is based on the coupled redox-complexation reaction between NAC, Fe(III) and TPTZ. The use of TPTZ as chromogenic reagent has improved selectivity, linearity and sensitivity of measurements. The method was successfully applied for determination of NAC in pharmaceutical formulations. The flow-injection manifold for spectrophotometric determination of NAC is showed in Figure 5.

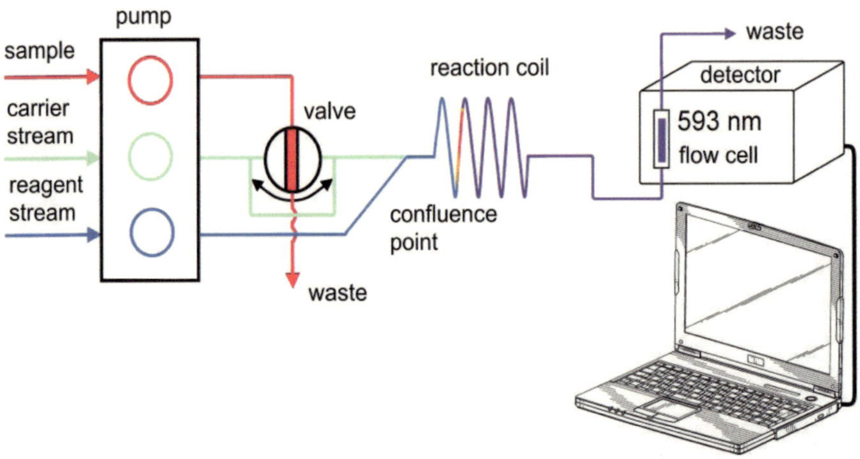

Figure 5. Flow-injection manifold configuration. Sample or standard solution (NAC); carrier stream (ultra pure water); reagent stream (1.0×10^{-3} mol L^{-1} Fe(III) and 1.0×10^{-3} mol L^{-1} TPTZ in acetate buffer solution, pH 3.6); peristaltic pump (flow rate 2.0 mL min^{-1}); injector valve (loop = 500 µL); confluence point (Y-type); reactor in coiled form (length: 300 cm, i.d. 0.8 mm); spectrophotometric detector (λ = 593 nm) equipped with flow cell (internal volume 160 µL).

In order to evaluate the potential of the proposed method for the analysis of real samples, flow-injection spectrometric procedure was applied to different pharmaceutical formulations (granules, syrup and dispersible tablets) for the determination of NAC. Recorded peaks refer to samples A, B and C are showed in the Figure 6.

A FIA spectrophotometric procedure for determination of *N*-(2-mercaptopropionyl)-glycine (MPG), tiopronin, has been proposed [36]. Determination was also based on the coupled redox-complexation reaction between MPG, Fe(III) and TPTZ. This coupled reaction was usefully used in development of the FIA method for determination of ascorbic acid in pharmaceutical preparations [37]. The proposed method is simple, inexpensive, does not involve any pre-treatment procedure and has a high sample analysis frequency.

5. Potential response of solid state potentiometric chemical sensors, theoretical approach and analytical teaching experiment

In this chapter a solid state potentiometric chemical sensors (PCS) used as detectors in the presented kinetic methods, performed in batch or flow-injection mode, are discussed. PCS make the use of the development of an electrical potential at the surface of a solid material when it is placed in a solution containing species which can be exchange (or reversibly react) with the surface. The species recognition process is achieved with a PCS through a chemical reaction at the sensor surface. Thus the sensor surface must contain a component which will react chemically and reversibly with the analyte in a contacting solution. The response of a solid state PCS to sensed ions in solution is governed by ion exchange or redox processes occurring between the electrode membrane and the solution. Since the transfer of the ions or electrons occurs across this membrane-solution interface, it is readily apparent that any changes in the nature and composition of the membrane surface will affect these processes and hence the response of the sensor. The potential of PCS in kinetic experiments is formed due to heterogeneous reaction at the surface of membrane and homogeneous reaction in contacting solution. The potential response of solid state PCS with Ag_2S + AgI membrane has been extensively investigated in our laboratory. For better understanding the behavior of this sensor in kinetic experiments the following questions are discussed. i) Which chemical compound on the surface of the membrane is important for the response of the sensor? ii)Which heterogeneous chemical reaction (or reactions), occurring between the electrode membrane and the sensed ions in solution, forms the interfacial potential? iii) Which homogeneous chemical reaction (reactions) in solution is (are) important for the potential response of the sensor? Potentiometric measurements with PCS containing membrane prepared by pressing sparingly soluble inorganic salts can be used for teaching homogeneous and heterogeneous equilibrium. Learning objective is to distinguish between homogeneous and heterogeneous equilibrium, and between single-component and multi-component systems [38, 39].

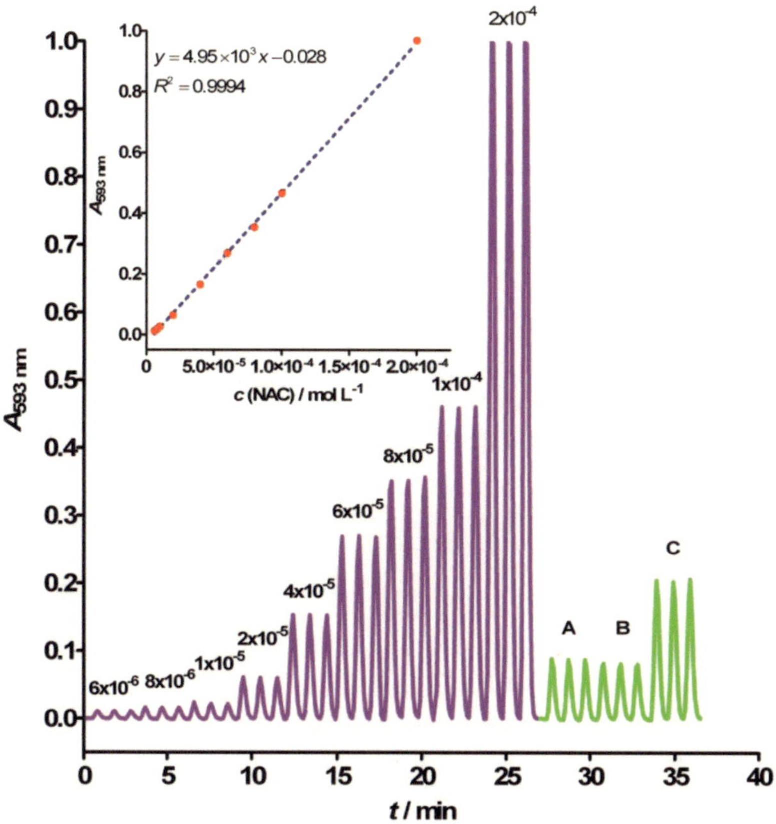

Figure 6. Fiagram chart and calibration curve (inlet) for spectrophotometric determination of NAC over the concentration range from 6.0 × 10^{-6} to 2.0 × 10^{-4} mol L^{-1}. Fiagram includes recorded peaks for three samples: (A) Fluimukan granules; (B) Fluimukan Akut Junior syrup and (C) Fluimukan Akut dispersible tablets

As it has been discussed, the determination of penicillamine was based on a batch and FIA experiments using PCS with AgI membrane. The membrane was prepared by pressing silver salts (AgI, Ag_2S) and powdered Teflon (PTFE). This AgI-based membrane detector, sensitive to sulfhydryl group, can be applied to flow-injection determination of different compounds containing sulfur. In order to understand the effect of stirring or flowing to potential response of sensor, for both kind of kinetic experiment (batch and FIA), it is necessary to develop a picture of liquid flow patterns near the surface of sensor in a stirred or flowing solution.

According to Skoog [40] three types of flow can be identified. *Turbulent flow* occurs in the bulk of the solution away from the electrode and can be considered only in stirred solution during batch kinetic experiment. Near the surface of electrode *laminar flow* take place. In FIA experiment only laminar flow exists in the tube. For both kind of kinetic experiments (batch and FIA) at 0.01 - 0.50 mm from the surface of electrode, the rate of laminar flow approaches zero and gives a very thin layer of stagnant solution, which is called the *Nernst diffusion layer* (Ndl). According Equation (13), the potential of sensor is determined by activity of Ag^+ ion in Ndl.

When the membrane of the sensor, containing both Ag_2S and AgI, is immersed in a solution with Ag^+ or I^- ions heterogeneous equilibrium at the phase boundary is established. The potential difference between the solution phase and the solid phase of the sensor is built up by a charge separation mechanism in which silver ions distribute across the membrane/solution interface as shown in Figure 7.

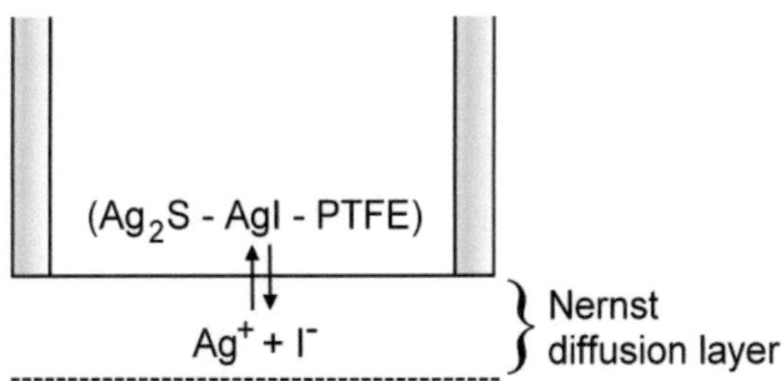

Figure 7. Heterogeneous equilibrium at the phase boundary between AgI-based membrane and solution.

The stable potential of PCS with AgI + Ag_2S membrane in contact with penicillamine (RSH) solution can be explained by the following consideration. According to the picture of liquid flow near the surface of sensor in a stirred or a flowing solution (Fig. 7) the potential of the sensor is determined by activity of Ag^+ ion in Ndl. In FIA experiment PCS with AgI + Ag_2S membrane (before injection of penicillamine) was in contact with flowing solution of Ag^+ ion, and the concentration of Ag^+ ions in solution including Ndl was 6.30×10^{-6} mol L^{-1}. The formation a new solid state phase in Ndl or/and at the surface of the sensing part of the tubular flow-through electrode unit may be expressed by the next reaction:

$$Ag^{+} + RSH \rightleftarrows RSAg(s) + H^{+} \tag{26}$$

with appropriate equilibrium constant.

$$K_{eq} = \frac{[H^{+}]}{[Ag^{+}]\cdot[RSH]} \tag{27}$$

$$K_{eq} = \frac{K_a}{K_{sp,RSAg}} \tag{28}$$

By using the experimentally established constant of solubility product [32] $K_{sp,\,RSAg}$, the dissociation constant of penicillamine, [41] K_a, and Equation (28) the equilibrium constant can be calculated.

$$K_{eq} = \frac{3.16\times10^{-11}}{1.40\times10^{-20}} = 2.26\times10^{9}$$

The calculated value of equilibrium constant suggests completeness of the new phase formation reaction at the surface of membrane. In addition, it can be supposed that, by adsorption process, both parts of membrane, AgI and Ag_2S, are covered with a thin layer of RSAg precipitate (Fig. 8). Under these conditions, the equilibrium activity of Ag^{+} ions and the corresponding response of PCS are governed by new heterogeneous equilibrium.

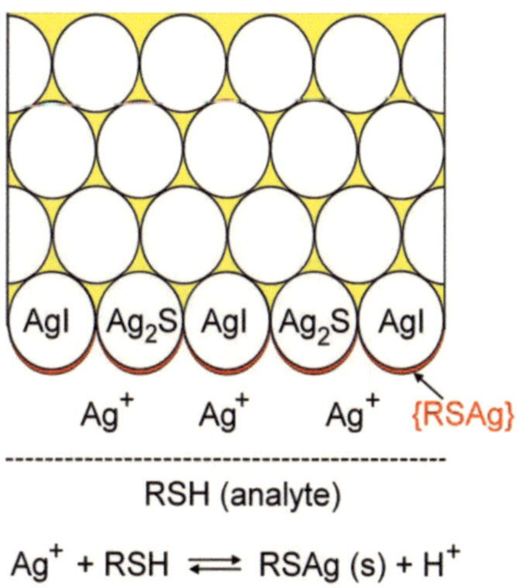

Figure 8. New phase formation in NdI and its adsorption on the surface of membrane.

Now we can calculate the minimal concentration of penicillamine, or any other RSH compound, which cause precipitation of RSAg in acid media.

$$K_{eq} = \frac{\left[H^+\right]}{\left[Ag^+\right] \cdot c(\mathrm{RSH}) \cdot \alpha(\mathrm{RSH})} = 2.26 \times 10^9 \tag{29}$$

$$c(\mathrm{RSH}) \geq \frac{\left[H^+\right]}{\left[Ag^+\right] \cdot K_{eq} \cdot \alpha(\mathrm{RSH})} \tag{30}$$

$$\alpha\left(\mathrm{RSH}\right) = \frac{\left[\mathrm{RSH}\right]}{c\left(\mathrm{RSH}\right)} = \frac{\left[\mathrm{RSH}\right]}{\left[\mathrm{RS}^-\right] + \left[\mathrm{RSH}\right]} \tag{31}$$

If we express $[RS^-]$ with dissociation constant of RSH,

$$\left[\mathrm{RS}^-\right] = \frac{K_{a,\,\mathrm{RSH}} \cdot \left[\mathrm{RSH}\right]}{\left[H^+\right]}$$

we obtain

$$\alpha\left(\mathrm{RSH}\right) = \frac{\left[H^+\right]}{K_{a,\,\mathrm{RSH}} + \left[H^+\right]} \tag{32}$$

In 0.100 mol L^{-1} perchloric acid, where experiment was performed, $\alpha(\mathrm{RSH}) \approx 1$.

$$c(\mathrm{RSH}) \geq \frac{0.100}{6.3 \times 10^{-6} \cdot 2.29 \times 10^9 \cdot 1}$$

$$c(\mathrm{RSH}) \geq 7.0 \times 10^{-6} \text{ mol L}^{-1}$$

This concentration of penicillamine may be estimated as the detection limit for describing experimental conditions. The solution of Ag^+ was pumped as a reagent in a two-line flow manifold typically at a concentration of 10^{-5} mol L^{-1} with 0.1 mol L^{-1} perchloric acid as a pH and ionic-strength adjuster.

List of abbreviations

AA	L-ascorbic acid
cys	L-cysteine
DA	dehydrogenized (oxidised) form of ascorbic acid
EDTA	ethylenediaminetetraacetic acid
FIA	flow injection analysis
FISE	fluoride ion-selective electrode

glu	L-glutatione
H_2A	reduced form of ascorbic acid
LOV	lab-on-valve
MPG	*N*-(2-mercaptopropionyl)-glycine
NAC	*N*-acetyl-L-cysteine
Ndl	Nernst diffusion layer
PCS	potentiometric chemical sensor
pen	D-penicillamine
phen	1,10-phenantroline
PTFE	polytetrafluoroethylene, Teflon
RSH	thiol compound
SIA	sequential injection analysis
TPTZ	2,4,6-trypyridyl-*s*-triazine

Author details

Njegomir Radić and Lea Kukoc-Modun
Department of Analytical Chemistry, Faculty of Chemistry and Technology, University of Split, Croatia

6. References

[1] Mottola HA. Kinetic aspects of analytical chemistry. New York: John Wiley & Sons 1988.

[2] Mark HB, Rechnitz GA. Kinetics in analytical chemistry. New York: John Wiley & Sons 1968.

[3] Cattrall RW. Chemical Sensors. Oxford: Oxford University Press 1997.

[4] Brand MJD, Rechnitz GA. Surface films on glass membrane electrodes [1]. Analytical Chemistry. 1970;42(2):304-305.

[5] Srinivasan K, Rechnitz GA. Reaction rate measurements with fluoride ion-selective membrane electrode: Formation kinetics of ferrous fluoride and aluminum fluoride complexes. Analytical Chemistry. 1968;40(12):1818-1825.

[6] Radić N, Komljenović J. Kinetic potentiometric determination of Fe(III) using s fluoride ion-selective electrode. Croatica Chemica Acta. 1991;64:679-687.

[7] Radić N. Determination of nanomole amounts of aluminium by use of a fluoride ion-selective electrode. The Analyst. 1976;101:657-660.

[8] Trojanowicz M, Hulanicki A. Microdetermination of aluminium with fluoride-selective electrode. Mikrochimica Acta. 1981;76(1-2):17-28.

[9] Radić N, Bralić M. Kinetic - potentiometric determination of aluminium in acidic solution using a fluoride ion-selective electrode. Analyst. 1990;115(6):737-739.

[10] Plankey BJ, Patterson HH, Cronan CS. Kinetics of aluminum fluoride complexation in acidic waters. Environmental Science and Technology. 1986;20(2):160-165.

[11] Radić N, Bralić M. Aluminium fluoride complexation and its ecological importance in the aquatic environment. Science of the Total Environment. 1995;172(2-3):237-243.

[12] Radić N, Vudrag M. Kinetic determination of iron(III) with a copper(II) selective electrode based on a metal displacement reaction. Journal of Electroanalytical Chemistry and Interfacial Electrochemistry. 1984;178(2):321-327.

[13] Martinović A, Radić N. Kinetic potentiometric determination of some thiols with iodide ion-sensitive electrode. Analytical Letters. 2007;40(15):2851-2859.

[14] Baumann EW, Wallace RM. Cupric-selective electrode with copper(II)-EDTA for end point detection in chelometric titrations of metal ions. Analytical Chemistry. 1969;41(14):2072-2073.

[15] Van Der Meer JM, Den Boef G, Van Der Linden WE. Solid-state ion-selective electrodes as end-point detectors in ompleximetric titrations. Part I. The titration of mixtures of two metals. Analytica Chimica Acta. 1975;76(2):261-268.

[16] Radić N. Metal (Fe3+, Cu2+) - EDTA complex formation studies in ethanol-aqueous mixtures with cupric ion-selective electrode. Analytical Letters. 1979;12:115 – 123.

[17] Radić N, Milišić M. Sulfide ion-selective electrode as potentiometric sensor for lead(II) ion in aqueous and nonaqueous medium. Analytical Letters. 1980;13:1013 – 1030.

[18] Midgley D, Torrance K. Potentiometric water analysis. 2 ed. New York: John Wiley & Sons 1991.

[19] Radić N, Prugo D, Bralić M. Potentiometric titration of aluminium with fluoride in organic solvent + water mixtures by use of a fluoride-selective electrode. Journal of Electroanalytical Chemistry. 1988;248(1):87-98.

[20] Radić N, Papa M. Electroanalytical Study of Aluminum Fluoride Complexation in 2-Propanol-Water Mixtures. Analytical Sciences. 1995;11:425-429.

[21] Martinović A, Kukoč-Modun L, Radić N. Kinetic spectrophotometric determination of thiols and ascorbic acid. Analytical Letters. 2007;40(4):805-815.

[22] Taha EA, Hassan NY, Abdel Aal F, Abdel Fattah LES. Kinetic spectrophotometic determination of acetylcysteine and carbocisteine in bulk powder and in drug formulations. ScienceAsia. 2008;34(1):107-113.

[23] Kukoc-Modun L, Radic N. Kinetic Spectrophotometric Determination of *N*-acetyl-L-cysteine Based on a Coupled Redox-Complexation Reaction. Analytical Sciences. 2010;26(4):491-495.

[24] Kukoc-Modun L, Radić N. Novel kinetic spectrophotometric method for determination of tiopronin [N-(2-mercaptopropionyl)-glycine]. Croatica Chemica Acta. 2010;83(2):189-195.

[25] Teshima N, Katsumata H, Kurihara M, Sakai T, Kawashima T. Flow-injection determination of copper(II) based on its catalysis on the redox reaction of cysteine with iron(III) in the presence of 1,10- phenanthroline. Talanta. 1999;50(1):41-47.

[26] Ruzicka J, Hansen EH. Flow injection analyses. Part I. A new concept of fast continuous flow analysis. Analytica Chimica Acta. 1975;78(1):145-157.

[27] Hansen EH, Miro M. How flow-injection analysis (FIA) over the past 25 years has changed our way of performing chemical analyses. TrAC - Trends in Analytical Chemistry. 2007;26(1):18-26.

[28] Van Staden JF. Flow injection determination of inorganic bromide in soils with a coated tubular solid-state bromide selective electrode. The Analyst. 1987;112(5):595-599,Toth K, Lindner E, Pungor E, Kolev SD. Flow-injection approach for the determination of the dynamic response characteristics of ion-selective electrodes. Part 2. Application to tubular solid-state iodide electrode. Analytica Chimica Acta. 1990;234(1):57-65.

[29] Komljenović J, Radić N. Design and properties of flow-through electrode with AgI-based membrane hydrophobized by PTFE; application to flow-injection determination of thiourea. Sensors and Actuators: B Chemical. 1995;24(1-3):312-316.

[30] Radić N, Komljenović J. Flow-injection determination of compounds containing sulfur by using potentiometric tubular sensor. Laboratory Robotics and Automation. 1998;10(3):143-149.

[31] Komljenović J, Radić N. Use of a multi-purpose solid-state ion-selective electrode body and an agl-based membrane hydrophobised by PTFE for the determination of i- and Hg2+. The Analyst. 1986;111(8):887-889.

[32] Radić N, Komljenović J, Dobčnik D. Determination of penicillamine by batch and flow-injection potentiometry with AgI-based sensor. Croatica Chemica Acta. 2000;73(1):263-277.

[33] Kolar M, Dobčnik D, Radić N. Potentiometric flow-injection determination of vitamin C and glutathione with a chemically prepared tubular silver electrode. Pharmazie. 2000;55(12):913-916,Kolar M, Dobčnik D, Radić N. Chemically treated silver electrodes for the determination of cysteine. Mikrochimica Acta. 2002;138(1-2):23-27.

[34] Bralić M, Radić N. Flow injection potentiometric determination of Fe(III) using a fluoride- selective electrode as detector. Analusis. 1999;27(1):57-60.

[35] Kukoc Modun L, Plazibat I, Radić N. Flow-injection spectrophotometric determination of *N*-acetyl-L-cysteine based on coupled redox-complexation reaction. Croatica Chemica Acta. 2011;84(1):81-86.

[36] Kukoc-Modun L, Radić N. Flow-injection spectrophotometric determination of tiopronin based on coupled redox-complexation reaction. Chemia Analityczna. 2009;54(5):871-882.

[37] Kukoc-Modun L, Biocic M, Radić N. Indirect method for spectrophotometric determination of ascorbic acid in pharmaceutical preparations with 2,4,6-tripyridyl-s-triazine by flow-injection analysis. Talanta. 2012;96:174-179.

[38] Radić N, Komljenović J. Potential response of solid state potentiometric chemical sensors, theoretical approach and teaching experiment. Acta Chimica Slovenica. 2005;52(4):450-454.

[39] Radić NJ, Dobčnik D. Surface compounds and reactions in relation to the response of solid state potentiometric chemical sensors. Surface Review and Letters. 2001;8(3-4):361-365.

[40] Skoog DA, West DM, Holler EJ, Crouch SR. Fundamentals of Analytical Chemistry. 8 ed. Belmont: Thomson 2004.

[41] The Merck Index, Merck Research Laboratories, New York, 1996.

Chapter 4

Analytical Chemistry Today and Tomorrow

Miguel Valcárcel

Additional information is available at the end of the chapter

http://dx.doi.org/10.5772/50497

1. Introduction

Dealing with Analytical Chemistry in isolation is a gross error [1]. In fact, real advances in Science and Technology —rather than redundancies with a low added value on similar topics— occur at interfaces, which are boundaries, crossroads —rather than barriers— between scientific and technical disciplines mutually profiting from their particular approaches and synergistic effects. Figure 1 depicts various types of interfaces involving Analytical Chemistry.

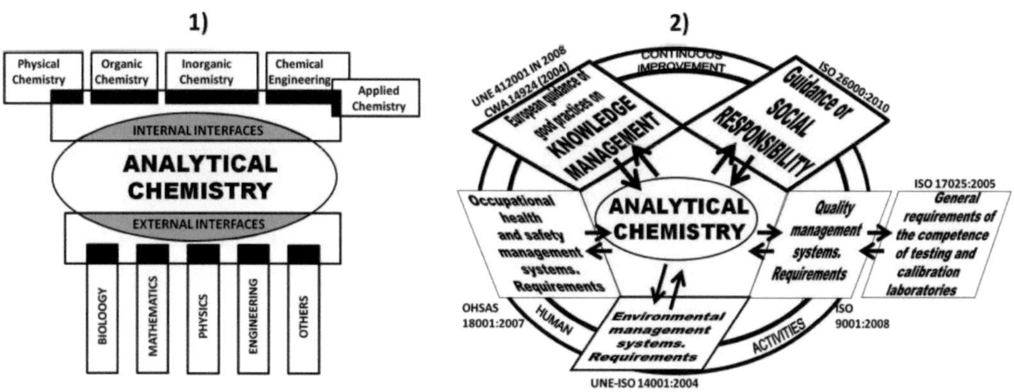

Figure 1. Analytical Chemistry at various interfaces. (1) Internal and interdisciplinary interfaces in the realm of Chemistry. (2) Interfaces with norms and guides. For details, see text.

Analytical Chemistry should in fact be present at a variety of interfaces such as those of Figure 1.1. Two belong to the realm of Chemistry (the framework of reference), namely:

1. *Internal interfaces* with other chemical areas (e.g. organic, inorganic, physical and applied chemistry, chemical engineering). Classifying Chemistry into these disciplines or subdisciplines, which are related via "fading" interfaces (1), has become obsolescent.
2. *External interfaces* with other scientific and technical disciplines such as biology, biochemistry, mathematics, physics or engineering, where Analytical Chemistry can play an active role (e.g. in the determination of enzyme activities or that of drugs of abuse in biological fluids) or a passive one (e.g. in chemometric developments for data processing or the use of immobilized enzymes in analytical processes).

Also, if Analytical Chemistry is to be coherent with its foundations, aims and objectives (see Section 2.2. of this chapter), it should establish two-way relationships with a variety of international written standards (norms and guides) in order to contribute to the continuous improvement of human activities (see Figure 1.2). The classical relationship between Analytical Chemistry and quality has materialized in ISO 17025:2005, which is the reference for laboratory accreditation. This norm contains technical requirements and other, management-related specifications that are shared with those in ISO 9001:2008, which is concerned with quality in general. Also, written standards dealing with knowledge management and social responsibility are highly relevant to the foundations and applications of Analytical Chemistry, even though they have rarely been considered jointly to date. In addition, Analytical Chemistry is very important for effective environmental protection, and occupational health and safety, since the (bio)chemical information it provides is crucial with a view to making correct decisions in these two complementary fields.

2. Cornerstones of modern analytical chemistry

Analytical Chemistry has evolved dramatically over the past few decades, from the traditional notion held for centuries to that of a modern, active discipline of Chemistry. Changes have revolved mainly around new ways of describing the discipline, and its aims and objectives, a broader notion of real basic references, the definition of the results of research and development activities and a holistic approach to analytical properties.

2.1. Definition

Analytical Chemistry can be defined in four simple ways as: (1) the discipline in charge of "Analysis" (the fourth component of Chemistry in addition to Theory, Synthesis and Applications, all of which are mutually related via the vertices of the tetrahedron in Figure 2); (2) the discipline in charge of the production of so named "(bio)chemical information" or "analytical information"; (3) the discipline of (bio)chemical measurements; and (4) the chemical metrological discipline, which is related to the previous definition.

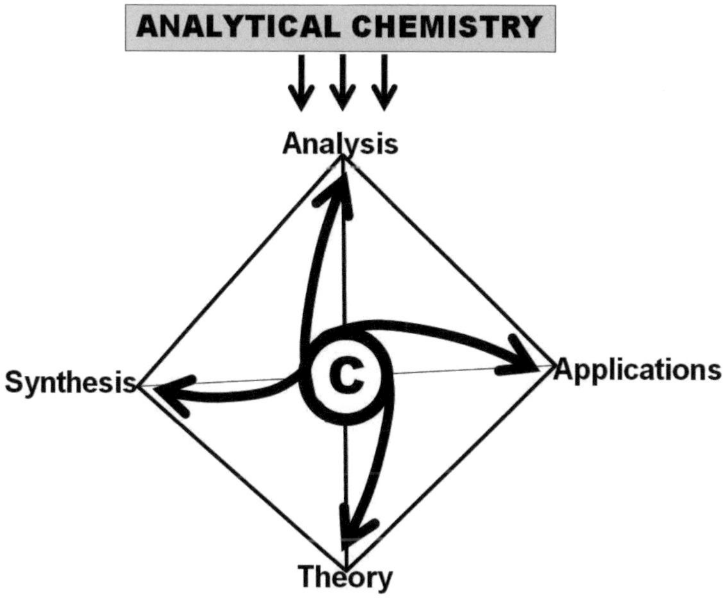

Figure 2. Analytical Chemistry is a discipline of Chemistry (C) inasmuch as it is responsible for "Analysis", an essential component of Chemistry in addition to theory, synthesis and applications in different fields (e.g. environmental science, agriculture, medicine).

These four general definitions have been used to formulate various more conventional definitions such as the following:

> *"Analytical Chemistry is a scientific discipline that develops and applies methods, instruments and strategies to obtain information on the composition and nature of matter in space and time"* (Working Party on Analytical Chemistry of the European Federation of Chemical Societies) [2].

> *"Analytical Chemistry is a metrological discipline that develops, optimizes and applies measurement processes intended to produce quality (bio)chemical information of global and partial type from natural and artificial objects and systems in order to solve analytical problems derived from information needs"* [3].

The strategic significance of Analytical Chemistry arises from the fact that it is an information discipline and, as such, essential to modern society. Analytical Chemistry as a scientific discipline has its own foundations, which materialize in keywords such as information, metrology, traceability, analytical properties, analytical problems and analytical measurement processes. Also, it shares some foundations with other scientific and technical areas such as Mathematics, Physics, Biology or Computer Science.

2.2. Aims and objectives

To be coherent with the previous definitions, Analytical Chemistry should have the aims and objectives depicted in Figure 3.

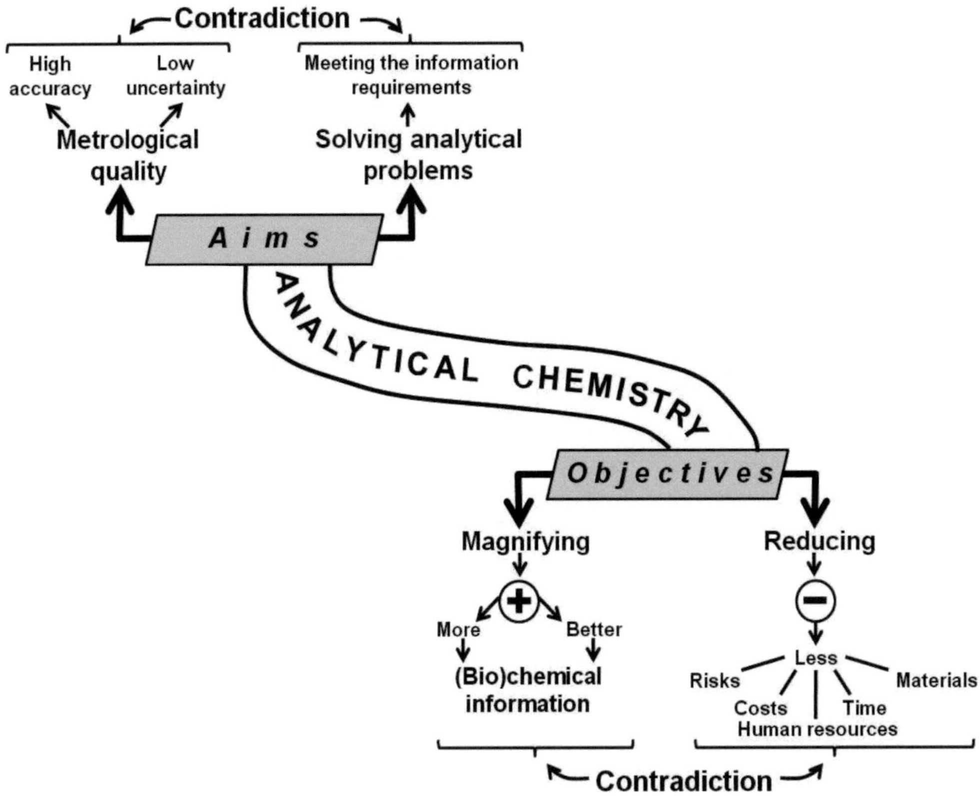

Figure 3. Primary aims and objectives of Analytical Chemistry. For details, see text.

Analytical Chemistry has two essential aims. One, which is intrinsic, is the obtainment of as high metrological quality as possible (i.e. of as true as possible analytical information with as low as possible uncertainty). The other, extrinsic aim is solving analytical problems derived from (bio)chemical information needs posed by "clients" engaged in a great variety of activities (health, general and agrifood industries, the environment).

The main magnifying objectives of Analytical Chemistry are to obtain a large amount of (bio)chemical information of a high quality, and its main reducing objectives to use less material (sample, reagents), time and human resources with minimal costs and risks for analysts and the environment.

The aims and objectives of Analytical Chemistry share its two sides (basic and applied); these are usually in contradiction and require appropriate harmonization. Thus, ensuring a high metrological quality may be incompatible with obtaining results in a rapid, economical manner. In fact, obtaining more, better (bio)chemical information usually requires spending more time, materials and human resources, as well as taking greater risks. Balancing the previous two aims and objectives requires adopting quality compromises [4] that should be clearly stated before specific analytical processes are selected and implemented.

2.3. Basic analytical standards

Analytical Chemistry relies on the three basic standards (milestones) shown in Figure 4 [5]. The two classical standards, which have been around for centuries, are tangible measurement standards (e.g. pure substances, certified reference materials) and written standards (e.g. the norms and guides of Figure 1, official and standard methods). A modern approach to Analytical Chemistry requires including a third standard: (bio)chemical information and its properties it should have to facilitate correct, timely decisions. Without this reference, analytical laboratory strategies and work make no sense. In fact, it is always essential to know the level of accuracy required, how rapidly the results are to be produced, and the maximum acceptable cost per sample (or analyte), among other requirements.

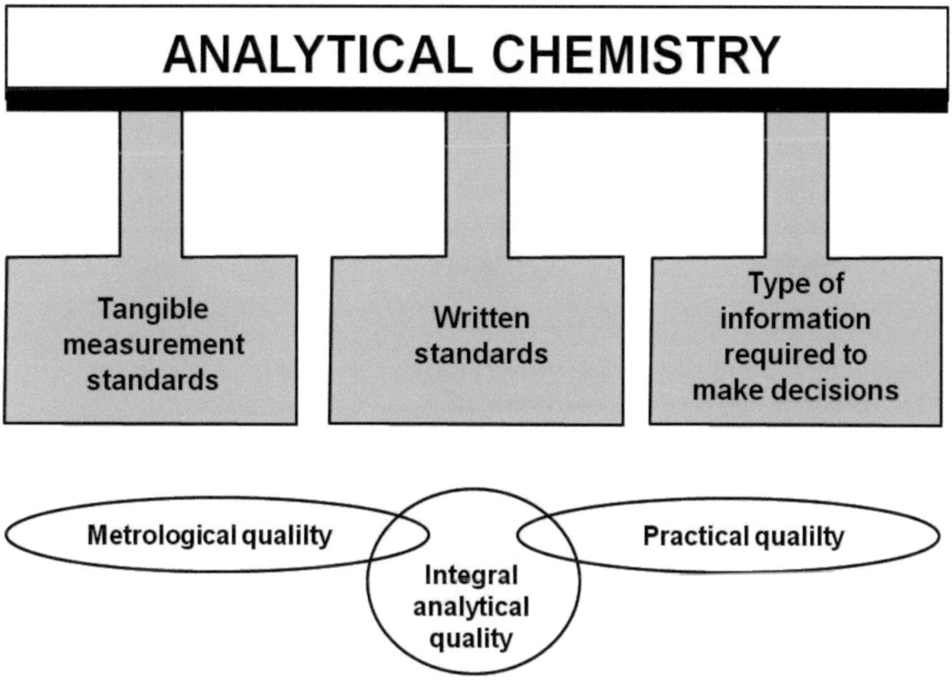

Figure 4. Basic standards supporting the Analytical Chemistry building and analytical quality related concepts. For details, see text.

As can be seen in Figure 4, conventional basic standards are related to so named "metrological quality", whereas (bio)chemical information and its required characteristics (the third basic standard) are related to "practical quality". Combining both concepts in so named "integral analytical quality" requires balancing two contradictory forces, which in turn entails the adoption of "quality compromises" (see Section 4 of this chapter).

2.4. R&D analytical "products"

The basic side of Analytical Chemistry encompasses a variety of R&D activities aimed at improving existing methods and/or developing new ones in response to new, challenging information needs. These activities can produce both tangible and intangible tools such as those of Figure 5 [6]. Typical tangible analytical tools include instruments, apparatus, certified reference materials, immobilized enzymes and engineering processes adapted to the laboratory scale (e.g. supercritical fluid extraction, freeze-drying). Analytical strategies, basic developments (e.g. calibration procedures) and chemometric approaches (e.g. new raw data treatments, experimental design of analytical methods) are the intangible outputs of analytical R&D activities. Transfer of technology in this context is more closely related to tangible R&D tools, whereas transfer of knowledge is mainly concerned with intangible R&D analytical tools; in any case, the two are difficult to distinguish.

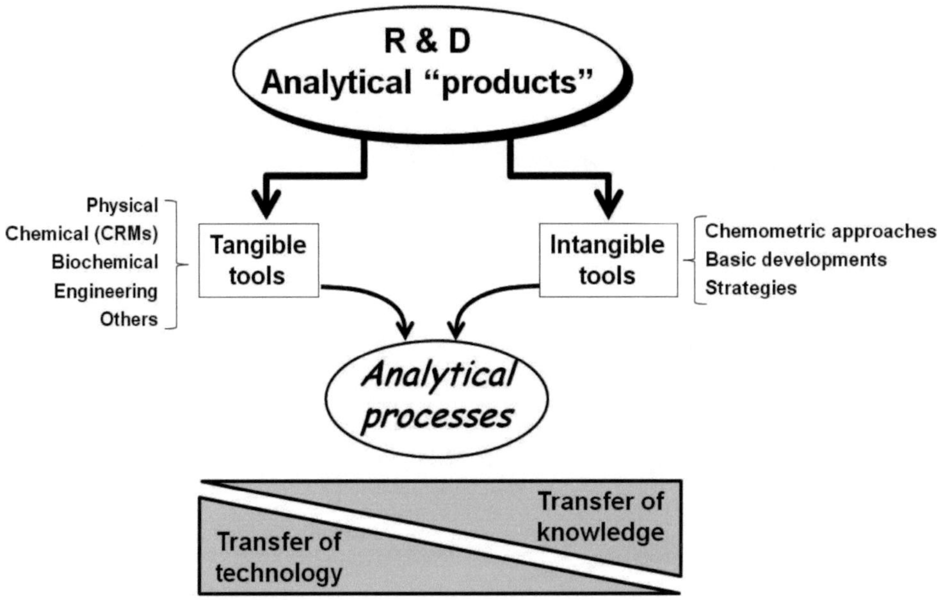

Figure 5. Main outputs of research and development (R&D) in Analytical Chemistry, transfer of knowledge and technology included. For details, see text.

2.5. Quality indicators

Analytical properties are quality indicators for the great variety of systems, tools and outputs of (bio)chemical processes that allow one to compare and validate analytical processes and the results they provide. Traditionally, they have been dealt with separately, with disregard of the high significance of their mutual relationships. Figure 6 provides a holistic view of analytical properties [7] as classified into three groups (capital, basic and productively-related) that are assigned to analytical results and analytical processes.

Top or capital analytical properties (accuracy and representativeness) are characteristics of the quantitative results of measurement processes. Accuracy is related to two classical metrological properties: traceability and uncertainty. In qualitative analysis, this property must be replaced with "reliability", which includes precision (a basic property). Capital properties can be defined in simple terms as follows:

Accuracy is the degree of consistency between a result (or the mean of several) and the true value or that considered as true (viz. the value for a certified reference material) in quantitative analyses. Any differences between the two constitute systematic errors.

Reliability is the proportion (percentage) of right yes/no answers provided by independent tests for analyte identification in aliquots of the same sample in qualitative analyses.

Representativeness is the degree of consistency of the results with the samples received by a laboratory, the overall sample or object studied, the particular analytical problem and the information required by the client.

Basic analytical properties (precision, robustness, sensibility selectivity) are attributes of analytical processes and provide support for capital properties. Thus, it is impossible to obtain highly accurate results if the analytical process is not precise, sensitive and selective enough. These properties can be defined as follows:

Precision is the degree of consistency among a set of results obtained by separately applying the same analytical method to individual aliquots of the same sample, the mean of the results constituting the reference for assessing deviations or random errors.

Robustness in an analytical method is the resistance to change in its results when applied to individual sample aliquots under slightly different experimental conditions.

Sensitivity is the ability of an analytical method to discriminate between samples containing a similar analyte concentration or, in other words, its ability to detect (qualitative analysis) or determine (quantitative analysis) small amounts of analyte in a sample.

Selectivity is the ability of an analytical method to produce qualitative or quantitative results exclusively dependent on the analytes present in the sample.

Productivity-related properties (expeditiousness, cost-effectiveness and personnel-related factors) are attributes of analytical processes with a very high practical relevance to most analytical problems.

Expeditiousness in an analytical method is its ability to rapidly develop the analytical process from raw sample to results. Expeditiousness is often expressed as the sample frequency (i.e. in samples per hour or per day).

Cost-effectiveness is the monetary cost of analyzing a sample with a given method and is commonly expressed as the price per analyte-sample pair. This property has two basic economic components, namely: the specific costs of using the required tools and the overhead costs of the laboratory performing the analyses.

Personnel-related factors. Strictly speaking, these are not analytical properties but are occasionally essential towards selecting an appropriate analytical method. These factors include the risks associated to the use of analytical tools and the analyst's safety and comfort.

As illustrated by Figure 6, quality in the results should go hand in hand with quality in the analytical process. In other words, capital analytical properties should rely on basic properties as their supports. It is a glaring error to deal with analytical properties in isolation as it has been usual for long. In fact, these properties are mutually related in ways that can be more consequential than the properties themselves. Their relationships are discussed in detail in Section 4. Each type of analytical problem has its own hierarchy of analytical properties, which materializes in the above-described "quality compromises".

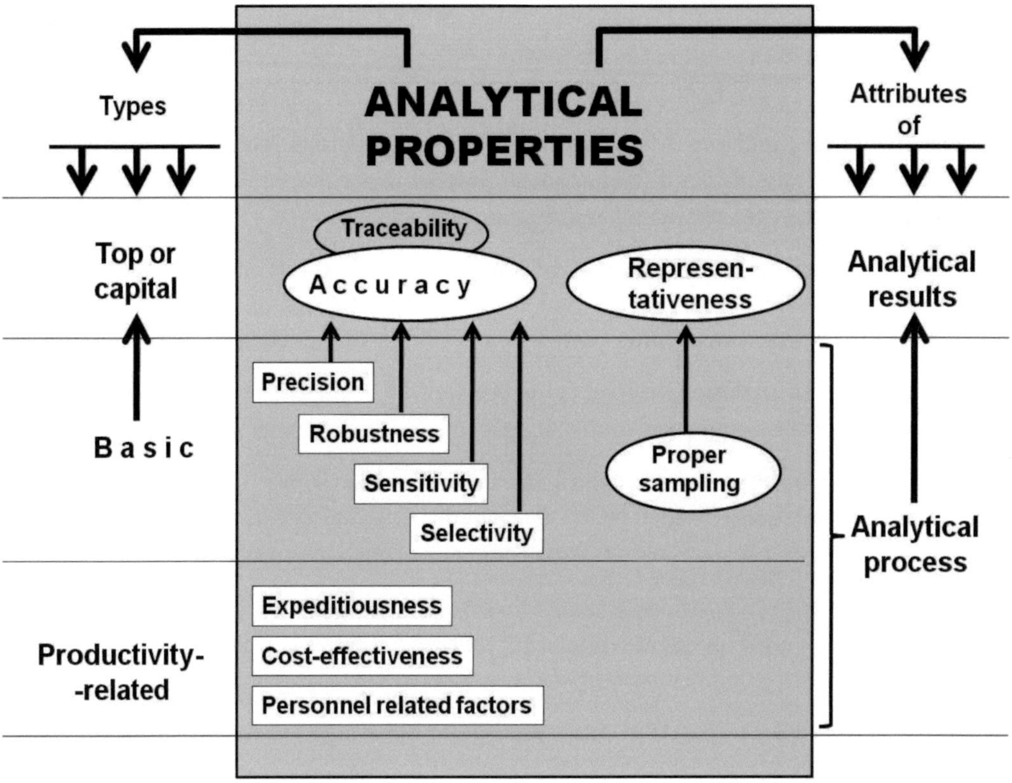

Figure 6. Holistic view of analytical properties as classified into three major groups and of their relationships with quality of the results and the analytical process. For details, see text.

3. (Bio)chemical information

The main output of (bio)chemical measurement processes is analytical or chemical/biochemical information, which is used to describe objects and systems for a variety of purposes, but especially to (*a*) understand processes and mechanisms in multidisciplinary approaches; and (*b*) provide support for grounded, efficient decision-making in a great variety of scientific, technical and economic fields. "Information" is probably the most important keyword for Analytical Chemistry, which has been aptly defined as an "information discipline" [8]. As shown below, (bio)chemical information lies in between raw data and knowledge; also, it has evolved markedly over the past few centuries and eventually become highly influential on human life and the environment by virtue of the increasing importance attached to social responsibility in Analytical Chemistry.

"(Bio)chemical information" and "analytical information" are two equivalent terms in practice. In fact, the difference between chemical and biochemical analysis is irrelevant as it depends on the nature of the analyte (e.g. sodium or proteins), sample (e.g. soil or human plasma) and tools involved (e.g. an organic reagent or immobilized enzymes).

3.1. Contextualization

Information is the link between raw data and knowledge in the hierarchical sequence of Figure 7. *Primary* or *raw data* are direct informative components of objects and/or systems, whereas *information* materializes in a detailed description of facts following compilation and processing of data, and *knowledge* is the result of contextualizing and discussing information in order to understand and interpret facts with a view to making grounded, timely decisions. Einstein [9] has proposed *imagination* as an additional step for the sequence in critical situations requiring the traditional boundaries of knowledge to be broken by establishing new paradigms.

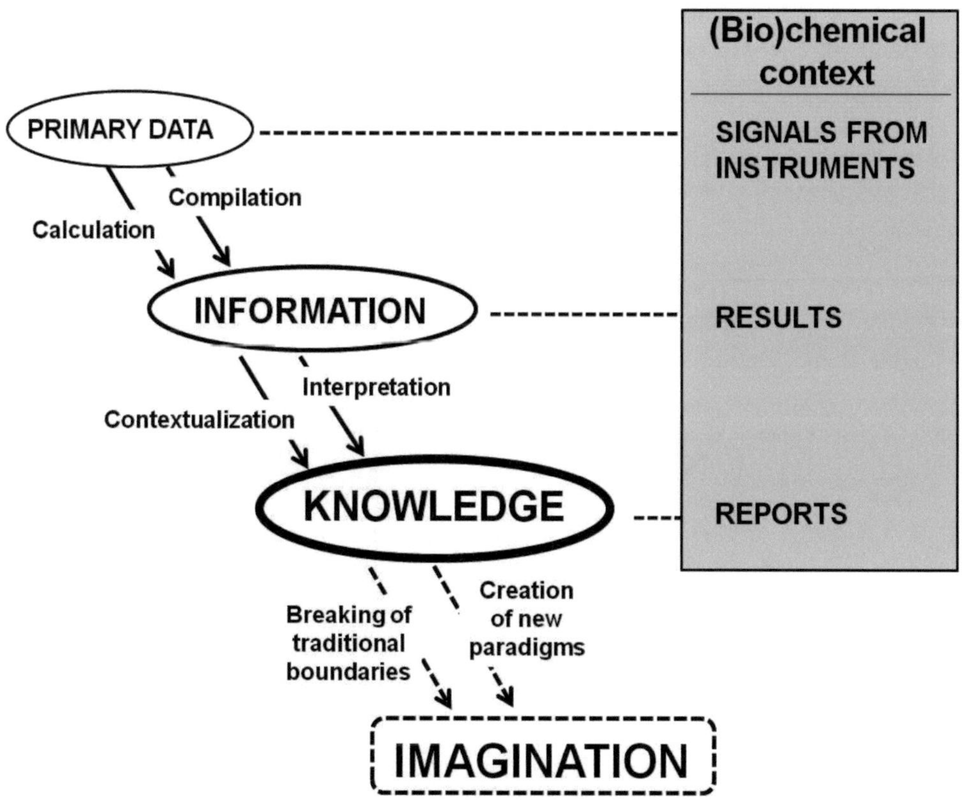

Figure 7. "Information" as an intermediate step between "raw data" and "knowledge", and their significance in the context of chemistry and biochemistry. For details, see text.

In a (bio)chemical context, "raw data" coincide with the primary "signals" provided by instruments (e.g. absorbance, fluorescence intensity, electrical potential readings). Also, "information" corresponds to the "results" of (bio)chemical measurement processes, which can be quantitative or qualitative. Finally, "knowledge" corresponds to "reports", which contextualize information, ensure consistency between the information required and that provided, and facilitate decision-making.

3.2. Types

Figure 8 shows several classifications of (bio)chemical information according to complementary criteria such as the relationship between the analyte(s) and result(s), the nature of the results, the required quality level in the results in relation to the analytical problem and the intrinsic quality of the results [10].

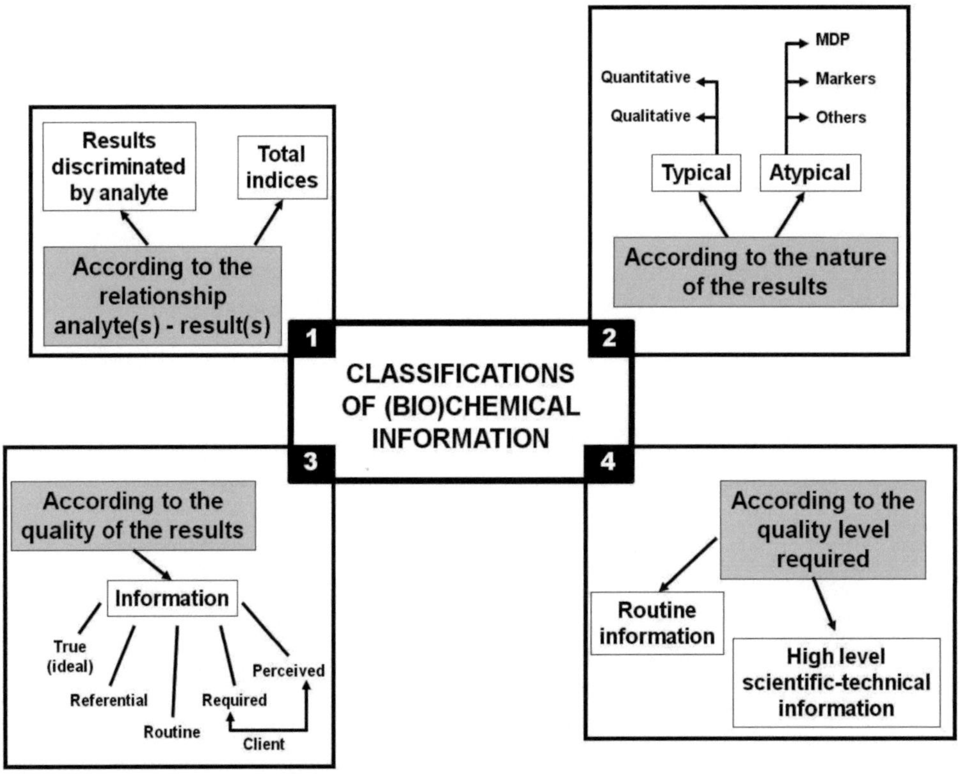

Figure 8. Four complementary classifications of (bio)chemical information based on different criteria. For details, see text.

Based on classification 1 in Figure 8, results can be discriminated by analyte (one analyte–one result), which is the most frequent situation when a separation (e.g. chromatographic, electrophoretic) is involved or when the measurement process is highly selective (e.g. immunoassays). Of increasing interest in this context are "total indices" [11], which can be defined as parameters representing a group of (bio)chemical species (analytes) having a similar structure/nature (e.g. greases, polyphenols, PAHs, PCBs) and/or exhibiting a similar operational behavior or effect (e.g. toxins, antioxidants, endocrine disruptors). More than 50% of the information required for decision-making is of this type. A large number of validated analytical methods produce this peculiar type of output. Probably, the greatest problem to be solved here is to obtain appropriate metrological support.

Classification 2 in Figure 8 establishes two types of results: typical and atypical. Typical (ordinary) results can be quantitative (viz. numerical data with an associated uncertainty range) and qualitative (e.g. yes/no binary responses); the latter have gained increasing importance in recent times. There are also atypical results requiring the use unconventional metrological approaches in response to specific social or economic problems. Thus, so named "method defined parameters" (MDPs) [12] are measurands that can only by obtained by using a specific analytical method —which, in fact, is the standard— and differ if another method is applied to the same sample to determine the same analyte. Usually, MDPs are total indices expressed in a quantitative manner (e.g. 0.4 mg/L total phenols in water; 0.02 mg/L total hydrocarbons in water). In some cases, MDPs are empirical (e.g. bitterness in beer or wine). Some can be converted into yes/no binary responses (e.g. to state whether a threshold limit imposed by legislation or the client has been exceeded). Markers [13] are especially important analytes in terms of information content (e.g. tumor markers, saliva markers to detect drug abuse).

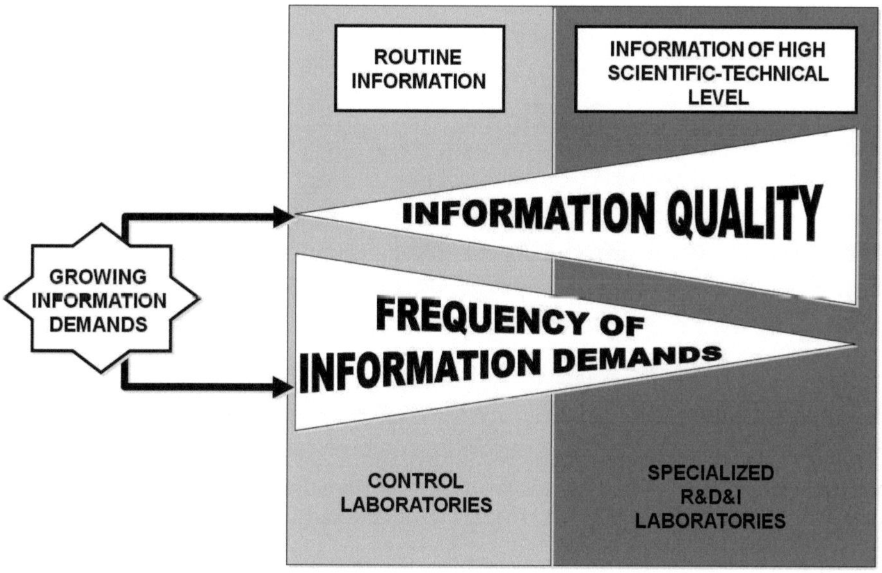

Figure 9. Contradiction between the frequency of information demands and the level of quality required in a situation of growing demands for (bio)chemical information. For details, see text.

Classification 3 in Figure 8 is based on the quality level of the results required in response to the client's information needs and comprises (*a*) routine information provided by control laboratories analyzing environmental, industrial, clinical or agrifood samples, for example; and (*b*) information of a high scientific and technical level that can only be obtained by using sophisticated instrumentation in specialized laboratories usually involved in R&D activities.

The frantic recent changes in social and economic activities have promoted an impressive expansion of (bio)chemical information about objects and systems. As can be seen in Figure 9, the quality of (bio)chemical information increases from routine laboratories to specialized laboratories, whereas the frequency of information demands decreases in the same direction. A compromise must often be made between these two contradictory notions. The panoramic view of Figure 9 is essential to perceive all connotations of analytical information. Classification 4 in Figure 8 is based on the intrinsic quality of the results and is examined in detail in Section 4 of this chapter.

3.3. Evolution

The routine information provided by control laboratories has evolved dramatically in the last decades. Figure 10 summarizes the most salient general trends in this context, which are commented on briefly below.

1. *Simplification.* Instead of delivering large amounts of high-quality information (a classical paradigm in Analytical Chemistry), there is a growing trend to delivering the information strictly required to make grounded decisions while avoiding time-consuming efforts to obtain oversized information that is useless in practice. Specially relevant here is the third basic standard supporting Analytical Chemistry (see Figure 4). The situation is quite common in routine laboratories but should be minimized or avoided altogether. Such is the case, for example, with the determination of hydrocarbons in tap water, the legal threshold limit for which is 0.1 ng/mL total hydrocarbons. Using a classical method involving several steps (e.g. filtration, cleanup, solvent changeover) and sophisticated equipment (e.g. a gas chromatograph and mass spectrometer) allows a long list of aliphatic and aromatic hydrocarbons with their concentrations —usually at the ppt or even lower level— to be produced which is utterly unnecessary to make grounded decisions, especially when a simplified method (e.g. one involving extraction into Cl_4C and FTIR measurement of the extract) can be used instead to obtain a total index totally fit for purpose.

2. *Binary responses.* Qualitative Analysis has been revitalized [14] by the increasing demand for this type of information; in fact, clients are now more interested in yes/no binary responses than in numerical data requiring discussion and interpretation. This trend is related to the previous one because obtaining a simple response usually entails using a simple testing method. The greatest challenge here is to ensure reliability in the absence of firm metrological support. In any case, false negatives should be avoided since they lead to premature termination of tests; by contrast, false positives can always and are commonly confirmed by using more sophisticated quantitative methodologies (see Section 5.4 and Figure 14).

3. *Total indices.* Based on classification 1 in Figure 8, a result can be a total index [11] representing a group of analytes having a common structure or behavior. This type of information is rather different from classical information, which is typically quantitative and discriminated by analyte. For example, the total antioxidant activity of a food can be easily determined with a simple, fast method using a commercially available dedicated analyzer. This avoids the usual

procedure for determining antioxidants in foodstuffs, which involves time-consuming sample treatment and the use of sophisticated instruments (e.g. a liquid chromatograph coupled to a mass spectrometer). This trend is also related to simplification and is rendering the classical paradigm of Analytical Chemistry (viz. maximizing selectivity) obsolete.

4. *Increasing importance of productivity-related properties.* The holistic approach to analytical properties of Figure 6, which considers hierarchical, complementary and contradictory relationships between them, and systematically using information needs as the third basic analytical reference (Figure 4), provide solid support for the increasingly popular productivity-related analytical properties (expeditiousness, cost-effectiveness and personnel-related factors). These properties are in contradiction with capital and basic analytical properties. Thus, achieving a high accuracy is not always the primary target and, in some cases, productivity-related properties are more important than capital properties. Such is the case with so named "point of care testing" approaches [15], the best known among which is that behind the glucose meter used to monitor the glucose level in blood at home. Glucose meter readings are inaccurate but rapid and convenient enough to control diabetes.

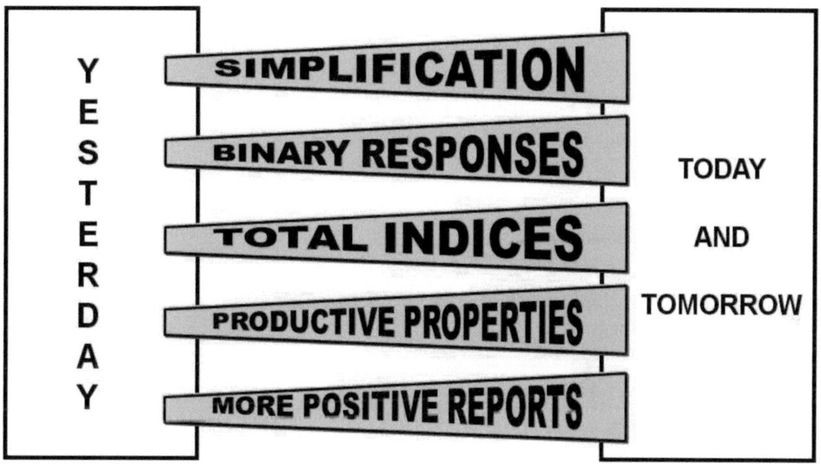

Figure 10. Major trends in the characteristics of (bio)chemical information provided by routine laboratories. For details, see text.

5. *Use of positive approaches to produce reports from results.* Analysts tend to emphasize negative aspects in delivering results and reports. A dramatic impulse of their "marketing abilities" to communicate with clients is therefore needed. One case in point is the word "uncertainty", inherited from Metrology in Physics and introduced in Metrology in Chemistry during the last few decades. This word can lead to wrong interpretations in chemistry nonmajors (e.g. politicians, economists, managers, judges) and raise global doubts about results. Simply replacing "uncertainty" with "confidence interval", which has the same scientific and technical meaning, can facilitate interpretation and acceptance of the results [16]. One other typical case is the use of "false positives" and "false negatives" to describe errors in binary responses. There is an obvious need to revise the terms related with (bio)chemical information and find alternatives emphasizing positive aspects rather negative connotations.

3.4. Social responsibility

Social responsibility (SR) is a concept encompassing a series of activities intended to support social well-being and help protect the environment which has extended from the corporate world to other human activities such as those involved in Science and Technology. In particular, Social Responsibility of Analytical Chemistry (SRAC) [17] is directed related to the impact of (bio)chemical information or knowledge from objects and systems to society, in general, and to human and animal health, the environment, industry and agrifoods, among others, in particular.

SRAC encompasses two basic requirements, namely: (1) producing reliable data, information and knowledge by using sustainable procedures in the framework of so named "green methods of analysis" [18]; and (2) ensuring consistency of delivered data, information and knowledge with the facts to avoid false expectations and unwarranted warnings.

Analytical Chemistry can therefore provide society with signals (data), results (information) and knowledge (reports), which can have a rather different impact. As can be seen in Figure 11, SRAC has two complementary connotations. One, intrinsic in nature, is the sustainable production of reliable data and results, and their appropriate transfer —which can be made difficult by contextualization and interpretation errors if left in the hands of nonexperts. The other, external connotation, is the appropriate delivery of reports (knowledge) to provide society with accurate information about the composition of natural and artificial objects and systems.

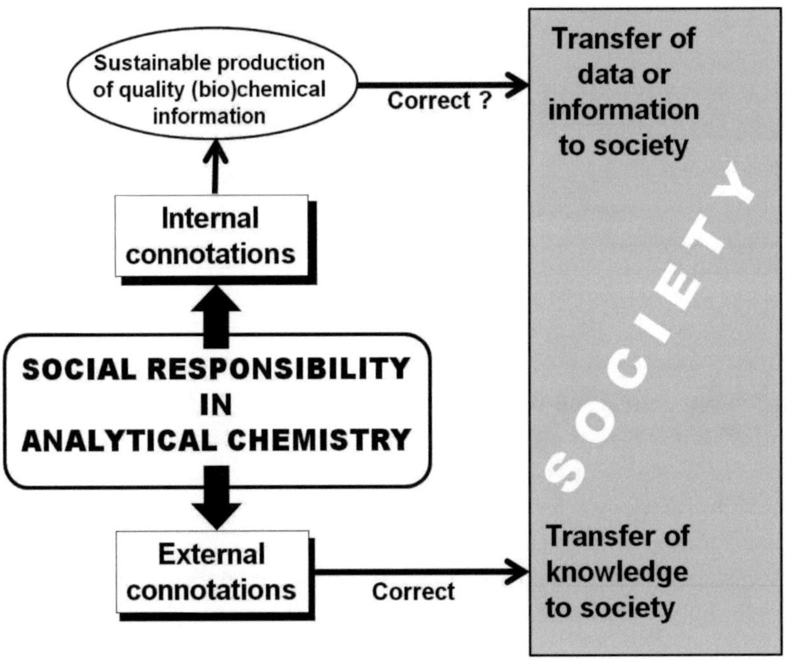

Figure 11. Connotations of Social Responsibility in Analytical Chemistry and ways to transfer data, information and knowledge to society. For details, see text.

4. Analytical quality

An integral approach to quality should rely on the following essential components: (1) the basic connotations of the concept as related to a set of features and comparisons, which in Analytical Chemistry materialize in analytical properties (Figure 6) and the three basic standards (Figure 4); (2) the practical connotations of fulfilling the (bio)chemical information needs posed by clients, which is one of the essential aims of Analytical Chemistry (Figure 3); and (3) the measurability of quality in terms of the capital, basic and productivity-related properties for analytical methods and their results.

Classification 3 in Figure 8 allows (bio)chemical information types to be depicted as shown in Figure 12, which additionally shows their mutual relationships via a tetrahedron. The arrows in the figure represent tendencies to converge —in the ideal situation, the tetrahedron could be replaced with a single, common point. Below is briefly described each member of the tetrahedron.

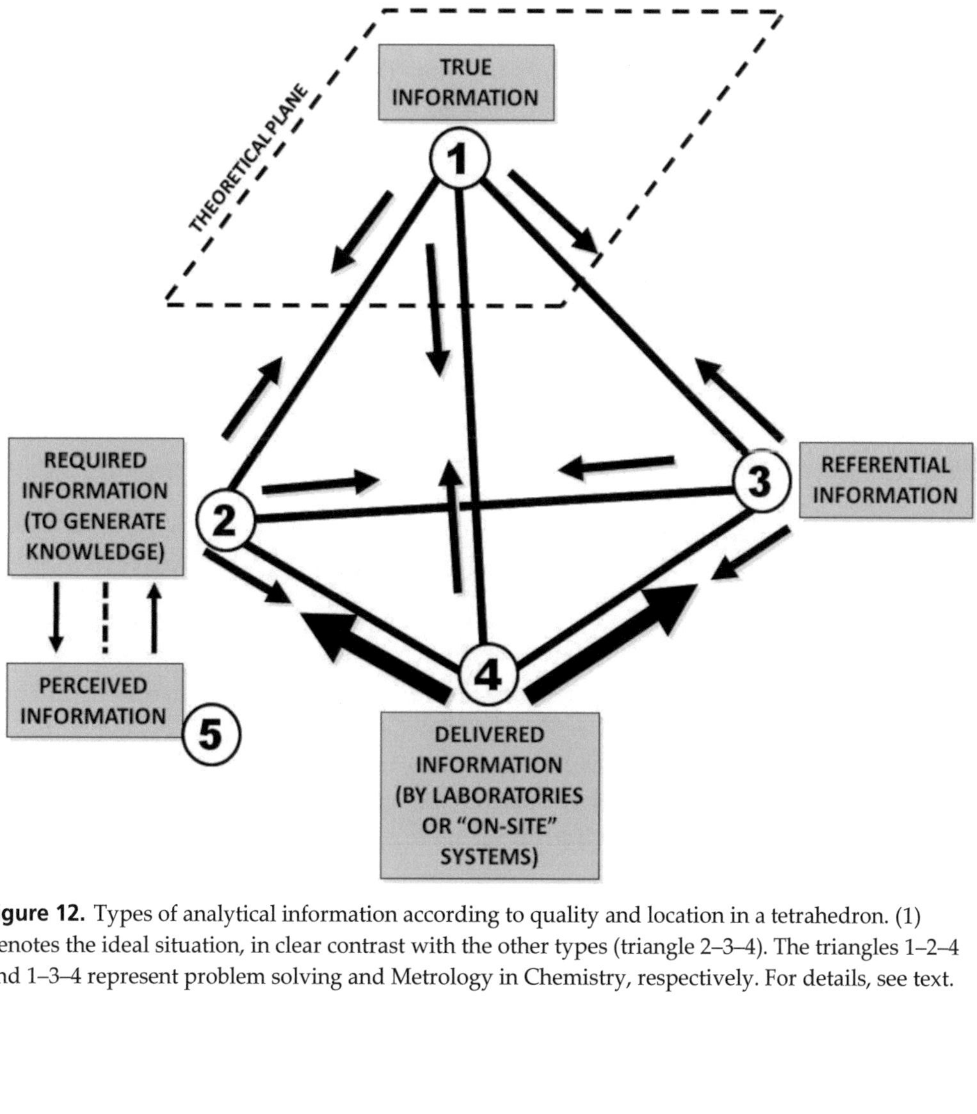

Figure 12. Types of analytical information according to quality and location in a tetrahedron. (1) denotes the ideal situation, in clear contrast with the other types (triangle 2–3–4). The triangles 1–2–4 and 1–3–4 represent problem solving and Metrology in Chemistry, respectively. For details, see text.

1. *True information* corresponds to intrinsic information about objects or systems. It is subject to no uncertainty and hence equivalent to trueness, which is unavailable to analysts. It is also known as "ideal analytical quality".

2. *Referential information* corresponds to the highest quality level that can be achieved in practice, with the information about a certified reference material (CRM) as the most typical example. Referential information is usually obtained in interlaboratory exercises where nonroutine laboratories analyze the same sample under the supervision of a renown organization (e.g. NIST in USA). Certified reference materials and their associated values are essential with a view to assuring quality in analytical methods and their results. The main problem here is their limited availability. In fact, only 3–5% of current needs for CRMs in (bio)chemical analysis have been met, in clear contrast with up to 90–95% in Metrology in Physics. Under these conditions, analysts are very often compelled to use alternative strategies to validate new analytical methods (e.g. standard addition procedures involving pure analytes).

3. *Routine information* is that produced by control laboratories or on-site systems operating outside the laboratory and largely used to control the quality of foodstuffs, industrial products or the environment.

4. *Required information* is that demanded by clients to make grounded, timely decisions and constitutes the third basic analytical standard (see Figure 4), which is frequently disregarded despite its high relevance to the major aims and objectives of Analytical Chemistry (see Figure 3).

5. *Perceived information,* which can be of a similar, higher or lower quality than that actually required by the client. Ideally, a client's perceived and required information should coincide. In some cases, the information delivered falls short of that required and can thus be deemed of low quality. Such is the case, for example, with the toxicological characterization of seawater by potential mercury contamination. The total mercury concentration is inadequate for this purpose because the toxicity of mercury species differs with their nature (inorganic, organometallic). It is therefore necessary to provide discriminate information for each potentially toxic mercury species.

The sides of the tetrahedron of Figure 12 represent the relationships between the different types of analytical information [19]. There are two contradictory relationships (forces) arising from delivered analytical information of great significance to Analytical Chemistry, namely: (1) the relationship between required and delivered information (2–4 in Figure 12), which represents problem solving and is related to the second aim of the discipline (see Figure 3); and (2) that between routinely delivered information and referential information (3–4 in Figure 12), which coincide at the highest metrological quality level —the first aim of Analytical Chemistry (Figure 3). One other significant distinction is that between required and perceived information on the client's side. Analytically, the most convenient situation is that where both types of information coincide in their level of quality —even though it is desirable that the client's perception surpass the actual requirements.

There are thus two contradictory facets of Analytical Chemistry that coincide with the its two aims, namely: a high level of metrological quality and fulfilling the client's information needs (see Figure 3). Analytical Chemistry is located at their interfaces [4]. There are some apparent conflicts, however, including (1) contradictory relationships of capital and basic analytical properties with productivity-related properties (see Figure 6); (2) failing to include required information among basic standards (see Figure 4); and (3) conceptual differences in analytical excellence between metrology and problem solving.

5. Major challenges

Achieving the general aims and objectives of Analytical chemistry in today's changing world requires producing tangible (reagents, sorbents, solvents, instruments, analyzers) and intangible means (strategies, calibration procedures, advances in basic science) to facilitate the development of new analytical methods or improvement of existing ones. This, however, is beyond the scope of this section, which is concerned with general trends in this context.

1. A sound balance between metrological and problem solving approaches for each information demand. The situation in each case depends strongly on the specific type of information and its characteristics (see Figure 8). With routine information, the challenge is to adopt well-defined quality compromises, which usually involves selecting and adapting analytical processes to fitness for purpose. Obtaining information of a higher scientific–technical level (e.g. that for materials used in R&D&I processes) calls for a high metrological quality level, as well as for exhaustive sample processing and sophisticated laboratory equipment.

2- Information required from objects/systems far from the ordinary macroscopic dimensions. These target objects or systems are directly inaccessible to humans because of their location or size. The size of such objects can fall at two very distant ends: nanomatter and outer space.

Analyzing the nanoworld is a real challenge for today's and tomorrow Analytical Chemistry. Extracting accurate information from nanostructured matter requires adopting a multidisciplinary approach. Nanotechnological information can be of three types according to nature; all are needed to properly describe and characterize nanomatter. Figure 13 shows the most salient types of physical, chemical and biological information that can be extracted from the nanoworld. Nanometrology, both physical and chemical, is still at an incipient stage of development. There is a current trend to using powerful hybrid instruments affording the almost simultaneous extraction of nanoinformation by using physical (e.g. atomic force microscopy, AFM) and chemical techniques (Raman and FTIR spectroscopies, electrochemistry).

The extraction of accurate information from objects and systems in outer space is a challenge at the other end of the "usual" range. This peculiar type of analysis uses miniaturized instruments requiring little maintenance and energy support. There are three different choices in this context, namely: (*a*) remote spectrometric analyses from spacecrafts with, for example, miniaturized X-ray spectrometers [20] or miniaturized mass spectrometers for the analysis of cosmic dust [21]; (*b*) analyses implemented by robots operating on the surface of other planets (e.g. to find traces of water in Mars [22], by using laser ionization-mass spectrometers [23]); or (*c*) monitoring of the inner and outer atmospheres of spacecrafts [24,25].

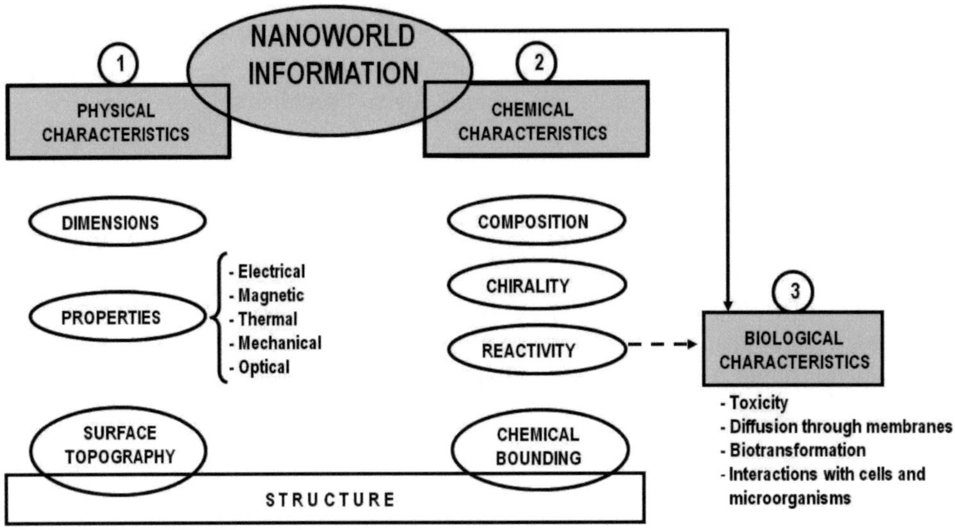

Figure 13. Types of information that can be extracted from the nanoworld. For details, see text.

3. *Breaking the traditional boundaries of the analytical laboratory.* To be consistent with its present aims and objectives (Figure 3), Analytical Chemistry cannot be exclusively confined inside the laboratory walls. In fact, it is necessary to open laboratory doors and analysts' minds in at least two complementary ways, namely:

(*a*) Analytical Chemistry should play an active role in activities preceding and following the development of analytical processes. Analytical chemists should play a twofold external role here by participating in the design and control of sampling procedures, and also in the discussion and interpretation of analytical results with other professionals in a multidisciplinary approach to transforming information (results) into knowledge (reports).

(*b*) Analytical Chemistry is increasingly focusing on the production of primary data from (automated) analytical processes implemented with so named "on site" systems outside the laboratory. These systems accumulate or send the requested primary data or results to a central laboratory. In the industrial field, on site monitoring can be performed "in-line" or "on-line". In clinical analysis, points of care testing systems (POCTs) [15] are extensively used for this purpose. The development of robust, reliable sensors for a broad range of analytes in a variety of sample types is a major challenge in this context, where automated calibration and quality control are the two greatest weaknesses.

4. *Vanguard–rearguard analytical strategies* [26]. As can be seen from Figure 9, the demand for (bio)chemical information has grown dramatically in the past decade and will continue to grow in the next. As a consequence, conventional analytical laboratories have been rendered unable to accurately process large numbers of samples each day. This has raised the need

for a new strategy (an intangible R&D&I analytical product according to Figure 5) intended to minimize the negative connotations of conventional sample treatment steps and facilitate the adoption of quality compromises between metrology and problem solving. This strategy uses a combination of vanguard (screening) systems and rearguard (conventional) systems as illustrated in Figure 14.

Vanguard analytical systems are in fact sample screening systems (SSS) [27,28] which are used in many activities where information is rapidly needed to make immediate decisions in relation to an analytical problem. Their most salient features are as follows: (*a*) simplicity (*viz.* the need for little or no sample treatment); (*b*) a low cost per sample–analyte pair; (*c*) a rapid response; (*d*) the production of atypical results (binary responses, total indices, method-defined parameters); and (*e*) reliability in the response. These systems act as mere sample filters or selectors and their greatest weakness is the low metrological quality of their responses —however, uncertainties up to 5–15% are usually accepted as a toll for rapidity and simplicity, which are essential and in contradiction with capital analytical properties. Sample screening systems provide a very attractive choice for solving analytical problems involving high frequency information demands. If these systems are to gain widespread, systematic use, they must overcome some barriers regarding accuracy (viz. the absence of false negatives for rapid binary responses), metrological support (traditionally, norms and guides have focused almost exclusively on quantitative data and their uncertainties) and commercial availability (e.g. in the form of dedicated instruments acting as analyzers for determining groups of analytes in a given type of sample such as antioxidants in foodstuffs or contaminants in water).

Rearguard analytical systems are those used to implement conventional analytical processes. Their most salient features are as follows: (*a*) they require conventional, preliminary operations for sample treatment and these involve intensive human participation and are difficult to automate (e.g. dissolution, solid and liquid extraction, solvent changeover); (*b*) they also usually require sophisticated instruments (e.g. GC–MS, GC–MS/MS, GC–FTIR/MS, LC–MS, LC–ICP-MS, CE–MS); (*c*) they afford high accuracy as a result of their excellent sensitivity and selectivity; (*d*) they use powerful primary data processing systems supported by massive databases easily containing 5000 to 50 000 spectra for pure substances, which ensures highly reliable results; (*e*) they usually provide information for each individual target analyte in isolation; and (*f*) they are expensive and operationally slow, but provide information of the highest possible quality level.

An appropriate combination of these two types of systems allows one to develop *vanguard–rearguard analytical strategies* (see Figure 14). With them, a large number of samples are subjected to the vanguard (screening) system to obtain binary or total index responses in a short time window. The output is named "crash results" and can be used to make immediate decisions. In fact, the vanguard system is used as a sample "filter" or selector to identify a given attribute in a reduced number of samples (e.g. a toxicity level exceeding the limit tolerated by law or by clients) which are subsequently processed systematically with the rearguard analytical system to obtain quantitative data and their uncertainty for each target analyte. The rich information thus obtained can be used for three complementary purposes, namely: (1) to confirm the crash results of vanguard systems (e.g. positives in binary responses to ensure that they are correct); (2) to amplify the simple (bio)chemical information provided by vanguard systems and convert global information about a group of analytes into discriminate information for each for purposes such as determining relative proportions; and (3) to check the quality of vanguard systems by using them to process a reduced number of randomly selected raw samples according to a systematic sampling plan.

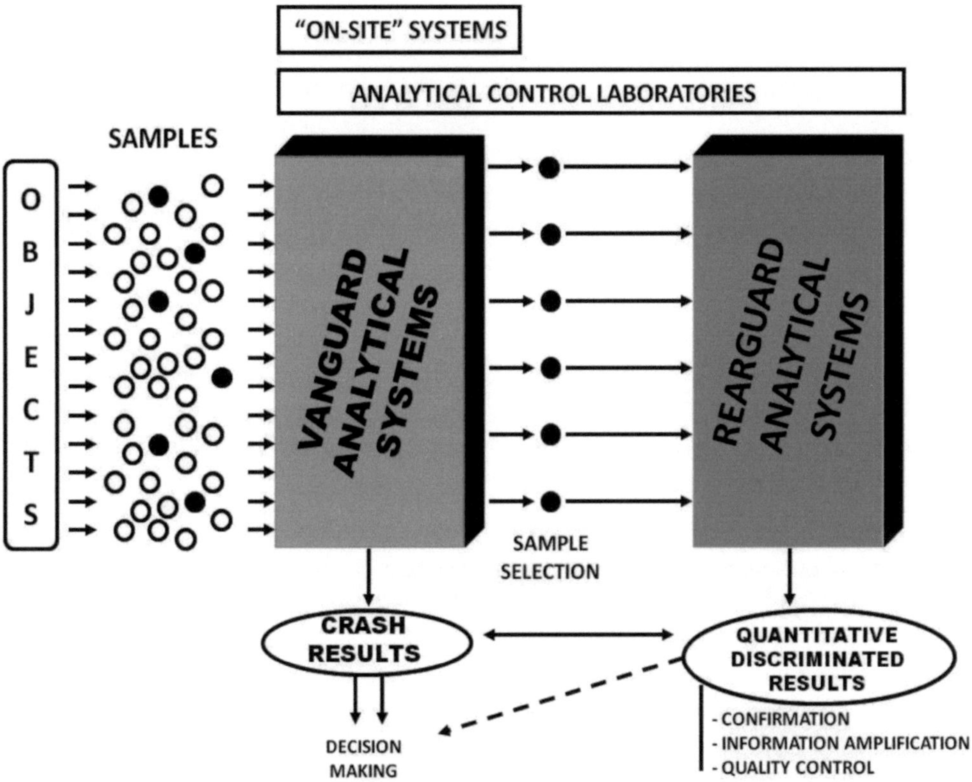

Figure 14. Vanguard–rearguard analytical strategies for the systematic analysis of large numbers of samples. For details, see text.

List of acronyms

R&D&I	Research, Development and Innovation
ISO	International Organization for Standardization
PAHs	Polycyclic Aromatic Hydrocarbons
PCBs	Polychlorinated Biphenyls
MDPs	Method Defined Parameters
FTIR	Fourier Transform Infrared Spectroscopy
SR	Social Responsibility

SRAC	Social Responsibility of Analytical Chemistry
CRM	Certified Reference Material
NIST	National Institute of Standards and Technology (USA)
AFM	Atomic Force Microscopy
POCTs	Point-of-Care-Testing
SSS	Sample Screening Systems
GC-MS	Gas Chromatography – Mass Spectrometry coupling
GC-MS/MS	Gas Chromatography – Mass Spectrometry / Mass Spectrometry coupling
GC-FTIR/MS	Gas Chromatography – Fourier Transform Infrared Spectroscopy / Mass Spectrometry coupling
LC-MS	Liquid Chromatography – Mass Spectrometry coupling
LC-ICP-MS	Liquid Chromatography – Inductively Coupled Plasma Spectrometry – Mass Spectrometry coupling
CE-MS	Capillary Electrophoresis – Mass Spectrometry coupling

Author details

Miguel Valcárcel
Faculty of Sciences of the University of Córdoba, Spain

Acknowledgement

The topic dealt with in this chapter was the subject of the author's lecture in his investiture as Doctor Honoris Causa by the University of Valencia (Spain) on March 30, 2011. This work has been supported by grant CTQ2011-23790 of the Spanish Government.

6. References

[1] R. Murray. "The permanency of fading boundaries" *Anal. Chem.*, 1996, 68, 457A.
[2] R. Kellner, J.M. Mermet, M. Otto, H.D. Widmer, M. Valcárcel "Analytical Chemistry" (2nd edition). 2004, *Wiley-VCH*, Weinheim, Germany.
[3] M. Valcárcel "Principles of Analytical Chemistry". 2000, *Springer–Verlag*, Heidelberg, pp 1–35.
[4] M. Valcárcel, B. Lendl "Analytical Chemistry at the interface between metrology and problem solving". *Trends Anal. Chem.* 2004, *23*, 527–534.
[5] M. Valcárcel, A. Ríos "Reliability of analytical information in the XXIst century". *Anal. Chim. Acta.* 1999, *400*, 425–432.
[6] M. Valcárcel, B.M. Simonet, S. Cárdenas "Bridging the gap between analytical R&D products and their use in practice". *Analyst.* 2007, *132*, 97–100.
[7] M. Valcárcel, A. Ríos "The hierarchy and relationships of analytical properties". *Anal. Chem.* 1993, *65*, 781A-787A.
[8] M. Valcárcel, E. Aguilera-Herrador "La información (bio)química de calidad". *An. Quím.* 2011, 107(1), 58–68.

[9] A. Einstein. La colección libre de citas y frases célebres en 1.4.E de http://es.wikiquote.org/wiki/Albert_Einstein.

[10] M. Valcárcel, B.M. Simonet "Types of analytical information and their mutual relationships". *Trends Anal. Chem.* 2008, *27*, 490–495.

[11] J.R. Baena, M. Gallego, M. Valcárcel. "Total indices in analytical science". *Trends Anal. Chem.* 2003, *22*, 641–646.

[12] B.M. Simonet, B. Lendl, M. Valcárcel "Method-defined parameters: measurands sometimes forgotten". *Trends Anal. Chem.* 2006, *25*, 520–527.

[13] J.R. Baena, M. Gallego, M. Valcárcel "Markers in Analytical Chemistry". *Trends Anal. Chem.*, 2002, *21*, 878–891.

[14] M. Valcárcel, S. Cárdenas, M. Gallego "Qualitative analysis revisited". *Crit. Rev. Anal. Chem.* 2000, *30*, 345–361.

[15] E. Aguilera-Herrador, M. Cruz-Vera, M. Valcárcel "Analytical connotations of point-of-care-testing". *Analyst* 2010, *135*, 2220–2232.

[16] J.D.R. Thomas "Reliability *versus* uncertainty for analytical measurements". *Ana*lyst. 1996, *121*, 1519.

[17] M. Valcárcel, R. Lucena "Social responsibility in Analytical Chemistry".*Trends Anal. Chem.*,2012, *31*, 1-7.

[18] S. Armenta, S. Garrigues, A de la Guardia "Green Analytical Chemistry". *Trends Anal. Chem.* 2008, *27*, 497–511.

[19] M. Valcárcel, A. Ríos "Required and delivered analytical information: the need for consistency". *Trends Anal. Chem.* 2000, *19*, 593–598.

[20] C.E. Schlemm et al. "The X-ray spectrometer on the Messenger spacecraft" *Space Sci. Rev.* 2007, *131*, 393–415.

[21] D.E. Austin, T.J. Ahrens, J.L. Beauchamp "Dustbuster: a compact impact-ionization time-of-flight mass spectrometer for in situ analysis of cosmic dust". *Rev. Sci. Instrum.* 2002, *73*, 185–189.

[22] E.K. Wilson "Mars watery mysteries". *C&E News (ACS)* 2008, December 1, pp 59–61.

[23] B. Sallé, J.L. Lacour, E. Vors, P. Fichet, S. Maurice, D.A. Cremers, R.S. Wiens "Laser-induced breakdown spectroscopy for mass surface analysis: capabilities at stand-off distances and detection of chlorine and sulfur elements". *Spectrochim. Acta B.* 2004, *59*, 1413–1422.

[24] M.L. Matney, S.W. Beck, T.F. Limero, J.T. James "Multisorbent tubes for collecting volatile organic compounds in spacecraft air". *AIHAJ* 2000, *61*, 69–75

[25] G.G. Rhoderick, W.J. Thor III, W.R. Miller Jr. F.R. Guenther, E. J. Gore, T.O. Fish "Gas standards development in support of NASA's sensor calibration program around the space shuttle". *Anal. Chem.* 2009, *81*, 3809–3815.

[26] M. Valcárcel, S. Cárdenas "Vanguard–rearguard analytical strategies". *Trends Anal. Chem.* 2005, *24*, 67–74.

[27] M. Valcárcel, S. Cárdenas, M. Gallego "Sample screening systems in Analytical Chemistry". *Trends Anal. Chem.* 1999, *18*, 685–694.

[28] M. Valcárcel, S. Cárdenas "Current and future screening systems". *Anal. Bioanal. Chem.* 2005, *381*, 81–83.

Chapter 5

Analytical Method Validation for Biopharmaceuticals

Izydor Apostol, Ira Krull and Drew Kelner

Additional information is available at the end of the chapter

http://dx.doi.org/10.5772/52561

1. Introduction

Method validation has a long and productive history in the pharmaceutical and now, biopharmaceutical industries, but it is an evolving discipline which changes with the times. Though much has been written about method validation for conventional, small molecule (SM) pharmaceuticals, less has appeared providing an overview of its application for complex, high molecular weight (MW) biopharmaceuticals (or biotechnology) products. This appears to be satisfyingly changing with the times, and this particular chapter has been designed to address this area of method validation. We hope to address herein the important issues of where do analytical method validation guidelines and directives stand today for biopharmaceutical (protein or related) products. Due to the recognized differences and complexity of biopharmaceuticals relative to small molecule drugs, regulatory agencies have accepted that what is expected of all SM, single molecule entities (even enantiomers), cannot be required for complex protein biopharmaceuticals, such as antibodies. While it is quite a simple matter, in most instances, to characterize and validate methods for SM drug substances, this is not always the case for complex biopharmaceuticals. Biotechnology products will always be heterogeneous mixtures of product-related species.

While the chapter below focuses on the principles and practice of method validation for biopharmaceuticals in the biotechnology industry, some comments on the topic of "academic method validation," and if and how that differs from what is required by the industry, seem warranted. In general, academics are not required by any regulatory agency or governmental body to perform any degree of method validation. However, one instance where it might be appropriate to do acceptable (whatever that means) method validation is when a reviewer of a grant proposal or manuscript destined for publication demands that some validation be performed. At times, Journal/Book Editors may suggest that some degree of method validation be performed, but in the final analysis, this requirement is at the discretion of reviewers. It appears that for the most part, little to no method validation is performed in academic circumstances, but on occasion, attempts are made to validate methods in academic laboratories. However, even there, such efforts do not begin to approach what is expected by regulatory agencies for industrial methods used to release a product for clinical trials use.

At times, in the past, Editors have taken the time to list what is expected in future submissions related to some degree of method validation. However, it was never obvious or clear that a lack of such studies really has ended up in manuscript rejections. Again, to a very large degree, this has depended on the rigorousness of the reviewers, resulting in somewhat a "luck of the draw" approach. Some may view this as frustrating and unfortunate, because in the absence of method validation, there should be no reason to accept the method and its applications, prima facie, or its results/data. However, academics somehow don't believe that method validation is required in order to do "good science" or publish. This situation has been changing for the better, but it is not quite where it really should be today. It will change when all editors, reviewers and manuscript/proposal submitters agree on the importance of doing good science by doing thorough and complete analytical method validation studies. Clearly, practitioners in the pharmaceutical and biopharmaceutical industries have much to offer to academic scientists in this regard.

One publication, years ago, appeared to demonstrate in certain, newer capillary electrochromatography (CEC) studies, unusually high plate counts and efficiencies. However, when others attempted to reproduce such results and data, nobody could come even close to what was in that original publication. Eventually, it was admitted that in the original study, none of those astounding results were reproducible or even replicable in a single lab. The work was never repeatable in their own hands, something that they conveniently forgot to mention anywhere in their papers. How could that happen? Well, it happened because neither the editors nor reviewers were thorough and rigorous in their demand for analytical method validation. They did not ask to see some evidence of repeatability, intermediate precision and other performance characteristics, a situation that would not be permissible in the industrial world due to regulatory requirements for method validation

Method validation in the pharmaceutical and biopharmaceutical industries is designed to help ensure patient safety during clinical trials and later when the drug becomes commercialized. While this reasoning is not applicable to basic research, and basic research in the academic community has at least one self correcting mechanism, peer verification, the lack of a requirement to document the performance characteristics of the methods in the academic world can, at times, lead to the publication of analytical methodologies, as noted above, that may lack scientific integrity.

2. Method validation for the biotechnology industry

The development of biotherapeutics is a complex, resource-intensive and time-consuming process, with approximately 10 years of effort from target validation to commercialization.

This reality, coupled with rapid technological advances and evolving regulatory expectations, impacts the ability of biotechnology companies to rapidly progress with development of their pipeline candidates.

Method validation is a critical activity in biopharmaceutical product development which often causes confusion and, at times, consternation on the part of analytical development teams. Questions surrounding method validation abound: (1) when should we validate our analytical methods? (2) what are the requirements for achieving method validation in a manner that is compliant across multiple regulatory jurisdictions around the world? (3) how can I implement a validation strategy that fits my company's business infrastructure and provides for seamless method transfer activities to other QC organizations in the company as well as contract QC organizations, when required?

Prior to proceeding to a discussion of method validation, it is important to differentiate amongst the categories of analytical methods used in the biopharmaceutical industry for product evaluation. In general, the analytical methods used can be divided into three categories: (1) screening methods; (2) release and stability methods; and (3) characterization methods. Screening methods are used to guide discovery research and process development. These methods, which are often carried out in high-throughput format using automation due to the large volumes of samples tested, do not typically follow any validation guidance, since they are not intended for a QC environment. Nonetheless, it is important to understand the capabilities and limitations of these methods so that the results can be appropriately applied to making decisions during process and product development. This is generally achieved through experience with the method in the analytical development organization. The second class of methods, release and stability methods, are intended for use in a Quality Control environment for product disposition and formal stability studies. In addition, these methods are sometimes used in QC for in-process samples in the form of in-process controls, which are used in the overall control strategy to ensure product quality (and for which the validation strategy should mirror that used for the release and stability methods). Whether used for release, stability, or in-process control applications, these methods are generally validated prior to the validation (conformance) lots to demonstrate that they have acceptable performance according to regulatory guidance (discussed below). The third class of methods, characterization methods, are used to support product characterization studies during reference standard characterization, process characterization, comparability studies, and other product characterization activities, and data from these studies is often submitted to regulatory agencies. Industry practice has recently evolved to meet regulatory expectations that these methods will be qualified according to written company procedures, though no formal written guidance is available, and method validation is not expected for these analytical procedures.

In order to meet current compliance expectations, an analytical method used to support GMP activities must be suitable for its intended use, and appropriate experimental work must be documented that provides this assurance. The demonstration of method suitability can be divided into two sets of activities: qualification and validation. When methods are new, under development, or subject to process or method changes, this activity is often called qualification, while more formal confirmation of method suitability for commercial applications is called validation (Ritter, Advant et al. 2004; Apostol and Kelner 2008; Apostol and Kelner 2008).

The strategy for method validation involves a continuum of activities that begins at the start of process and product development and carries through to the marketing application and beyond. Typically, analytical method development begins after the biological target has been identified and verified, the protein therapeutic has been defined (primary sequence), and the sponsor has made the decision to develop a manufacturing process that will enable human clinical trials. The initial demonstration that the method is suitable for its intended purpose for use as a release and stability method is generally carried out in the form of method qualification, an activity that generally takes place prior to the release of the material for first-in-human (Phase 1) clinical trials. At the later stage of product development, typically prior to the start of pivotal phase III clinical trials, method developers perform qualification studies which will enable method validation. Finally, method validation generally takes place prior to the release and stability testing of the validation manufacturing lots.

It should be noted that although method qualification, which evaluates the performance characteristics of the method against meaningful target expectations, is a critical development activity that establishes the suitability of the method for release of early to mid-phase clinical materials, this activity is not, to the best of our knowledge, clearly defined in regulatory guidance, which tends to focus on method validation. It is therefore difficult to define the scope of method qualification, though regulatory expectations and industry practices have evolved to define method qualification as a means to assure acceptable method performance during process and product development, prior to the formal validation exercise that occurs before the testing of the validation lots.

The necessity of method validation has been reinforced by a variety of national and international regulations (USP 1994; USP 1999; CDER 2001; ICH 2005) which are subject to user interpretation. For example, current GMP regulations, [21 CFR 211.194 (a)] require that methods used in testing of the samples meet proper standards of accuracy and repeatability. Validation provides assurance that this regulation is met. USP <1225> defines validation of analytical procedures as the process by which it is established by laboratory studies that the performance characteristics of the procedure meet the requirements of the intended analytical application. ICH guideline Q2R1 defines validation of analytical procedures as the demonstration that the method is suitable for its intended purpose. ICH guidance specifies that validation of analytical procedures needs to be included as part of the registration package submitted within the EU, Japan and USA. While the biotechnology industry, in a manner analogous to the pharmaceutical industry, is heavily regulated, the majority of the regulations are targeted at commercial products, leaving a significant gap in available regulatory guidance for earlier stages of product development. While numerous articles have been published to provide the scientific principles and exemplify the types of associated activities relevant for method validation (Swartz and Krull 1997; Shabir 2003; Ritter, Advant et al. 2004; ICH 2006; Krull and Swartz 2006; Swartz and Krull 2006; Swartz and Krull 2009), the most frequently referenced document, is the ICH guideline Q2R1, "Validation of analytical methods: text and methodology"(ICH 2005), This document covers validation activities targeted at product registration; hence, this guidance is specifically applicable to commercial products. Method qualification has emerged as the typical means of filling the gap for assessing the suitability of analytical method performance at earlier stages of product development.

3. Qualification of characterization methods

Characterization methods typically involve highly specialized technologies which are labor intensive and difficult to perform on a routine basis, which includes, for example, AUC, CD, FTIR, DSC, SEC-LS , and NMR. These methods are often used to supplement lot release methods to provide orthogonal detection/separation modes and/or to verify structural integrity (e.g. primary, secondary, tertiary structure). This is in contrast to Quality Control methods, which typically employ proven technologies to enable in-process controls, lot disposition and GMP stability assessment in the GMP laboratory setting, requiring stringent assessment of performance characteristics that follow ICH guidelines. Therefore, it is important to define an appropriate level of qualification for these complex and non-routine characterization methods. Industry practice has evolved multiple means of defining a qualification path for characterization methods, including:

- Ensuring the adherence to written technical procedures
- Ensuring that the equipment has a documented record of initial equipment qualification (IQ/OQ), preventative maintenance (PM), and/or calibration.
- Ensuring that data are generated by scientists with appropriate technical skills documented through training records and/or academic credentials.
- Ensuring that all experiments are accompanied by proper controls to ensure that the method is capable of measuring the intended attributes of the product. Control experiments should be designed in such a way that the quantitative aspect of the measurement can be clearly demonstrated from the results of the experiments. Properly controlled experiments should be performed to address the precision (repeatability) of the measurements.

Recently Jiang et al. provided an excellent review of the qualification of the biophysical methods including AUC, CD, FTIR, DSC, SEC-LS, MFI and LO based methods. The authors describe how qualification of these methods enables better knowledge of the methods and objective interpretation of the results. The general considerations described there can be applied to other biophysical methods as appropriate as well (Jiang, Li et al. 2012). In most cases qualification of biophysical methods is focused on the determination of precision and demonstration that the methods are suitable for their intended applications. Successful qualification enables the understanding of the method capability and the consistent determination of product attributes.

4. Qualification and validation of release and stability methods

Qualification should be performed prior to method implementation in the Quality laboratories to ensure the integrity of the data provided on the Certificate of Analysis for clinical lots. In most cases, at early stages of clinical development, only one sample type requires qualification because, in general, the drug substance (DS) and drug product (DP), for which specifications are established requiring testing in the Quality labs, often have the same composition (formulation). If this is not the case, a technical assessment should be made of whether differences in the matrix have the potential to impact the qualification results and, if so, a strategy for verifying the qualification status of the two sample types, relative to each other, should be devised. For example, full qualification of the DS can be followed by a matrix verification for the DP, generally in the form of a repeat of the specificity and precision evaluation.

Method validation is typically completed before process validation in adherence with cGMP procedures outlined in ICH Q2R1(ICH 2005) . Method validation for release and stability methods can be considered as the pivotal point in the method lifecycle because it justifies the use of the method in commercial settings to guide decisions about product disposition and lot stability. In addition, the validation activity provides a defined point of transfer of ownership of the methods from the development organization to the commercial (operations) organization. Typically, these activities are initiated after the sponsor has made a commitment to commercialize the drug candidate (which generally occurs after positive feedback from clinical trials).

ICH Q2R1 specifies that method validation has three components: assessment of performance characteristics, demonstration of robustness and system suitability. It should be noted that industry practice dictates that method qualification also evaluates these three components, with a noticeable difference, in that while validation has a formal protocol and pre-defined acceptance criteria for the performance characteristics, method qualification does not. It is a good practice to adopt general target expectations for method qualification as a means of evaluating the outcome of exploratory work on performance characteristics. The expectation should reflect the desired characteristics for the methods with respect to precision, range, QL, etc. If the method does not meet these expectations, the method should be re-developed and/or optimized. Recently, many companies have adopted the practice of developing and qualifying multiproduct methods that can be used for more than one product within specific molecular classes, such as monoclonal antibodies. In such instances, verification of performance could be adequate instead of full qualification studies once the method has undergone full qualification for the first molecule of the specified class.

Standard industry practice dictates that methods used to assess drug substance and drug product stability should show that they are able to detect changes in quality attributes. This can be demonstrated in forced degradation studies on the appropriate sample types using conditions known to impact protein quality, such as elevated temperature, pH extremes, and incubation with oxidizing agents such as hydrogen peroxide to induce molecular changes such as aggregation, deamidation, peptide bond cleavage and protein oxidation.

In addition to the stability indicating properties of the method, the assessment of sample stability can be considered as a pre-requisite to method validation. Sample stability can be divided into two activities – an evaluation of sample storage conditions prior to analysis, and assessment of the stability of prepared samples while waiting for analysis. Samples are often stored frozen after collection and thawed prior to analysis; in these instances, sample integrity should be assessed over a minimum of one freeze-thaw cycle for each sample type (preferably more than one cycle in most cases). Sample stability after preparation and before analysis (e.g., time spent in an auto-sampler) should be evaluated to determine the maximum duration of an assay (sequence). Details of sample handling should be included in the validation protocol.

Method validation confirms the performance characteristics demonstrated during method qualification and demonstrates the suitability of the method for commercial use. This confirmation effort should follow a pre-approved protocol with clear and justifiable acceptance criteria. In the context of the analytical lifecycle, the key components of method validation are as follows:

1. The experimental design of method validation should mimic the qualification design, and acceptance criteria should be linked to the target expectations used in the qualification experiments. In the absence of such rigor, validation experiments become exploratory research and run the risk of undermining the results of the method qualification.
2. Similarly to the qualification (in most cases), the validation acceptance criteria are set based on the type of method and should not differ from target expectations. When qualification target expectations are not met during a qualification study, the rationale for re-evaluation of the acceptance criteria should be proposed in the qualification summary. Setting acceptance criteria for the precision of a method frequently causes confusion, anxiety, and inconsistency in practice. For validation studies, requirements for a reporting interval aligned with the specification for precision studies provide excellent guidance for setting the acceptance criteria for precision and other performance characteristics.
3. Validation acceptance criteria should only include the objective parameters from the qualification to avoid any subjective interpretations, which could impact the outcome of the confirmatory validation studies. For example, frequently during qualification studies, scientists expect that the residual from linear regression does not show any bias (trend). Since the community has not adopted a uniform measure of the bias (which is frequently based on visual evaluation), it is not advisable to include such a requirement in the validation acceptance criteria.

Validation studies should be executed for sample types that will be routinely tested in GMP environments to make decisions about product disposition. This typically includes the following sample types:

- Sample types listed on all release and stability specifications (intermediates, drug substance and drug product);
- Samples associated with process controls and in-process decision points.

5. Performance characteristics

In order to produce a reliable assessment of method performance, all necessary performance characteristics should be evaluated in carefully designed experiments. ICH guideline Q2R1 specifies which performance characteristics should be evaluated for validation. However, interpretation of the table for protein products is not straightforward. This is due in part to the fact that ICH Q2R1, Q6B and the industry used different nomenclature to describe the type of methods. The table below details the performance characteristics that should be assessed during qualification, and subsequently during the confirmatory validation experiments for protein products.

ICH Q2R1 method types	ICH Q6B method types	Industry method types	Performance characteristics					
			Specificity	Linearity	Range	Precision	Accuracy	LOD/LOQ
Testing for impurities	Quantity	Titer	√	√	√	√	√	*
	Purity and impurities	Purity	√	√	√	√	√	√
		Immuno	√	√	√	√	√	√
		DNA	√	√	√	√	√	√
		Peptide map	√			√		√
		Gels	√	√	√	√	√	√
		Process Reagents	√	√	√	√	√	
Assay	Potency	Potency	√	√	√	√	√	√
Identification	Identity	ID	√					

* In some cases may be required by USP <1225>.

Table 1. Performance Characteristics that need to be Evaluated During Qualification/Validation by Method Type

6. Precision

Precision has typically been considered as the most important performance characteristic of the method, because it gives customers/clients of the analytical data direct information on the significance or uncertainly of results. Typically, method precision is established from replicate analyses of the same sample. However, methods for predicting precision have recently been published that allow the assessment of precision based on a single chromatogram (Apostol, Kelner et al. 2012).

Method precision defines the capability of the method expressed in its reporting interval (Holme and Peck 1998). Agut et al. (Agut, Segalini et al. 2006) examined different rules and their application to the reporting interval of results and specifications. The best known and simplest rule to implement is that stated in the AMST standard E-29-02. The rule states that the results of analytical measurements should be rounded to not less than 1/20 of the determined standard deviation (ASTM 2005).

For example, bioassays with a standard deviation of 11.8 should adopt a reporting interval larger than 0.59. However, this 0.59 reporting interval is impractical in day-to-day applications due to the inability of bioassays to provide precision that would justify such a reporting interval. Therefore, bioassays with a standard deviation of 11.8 would result in a reporting interval of 1. Similarly, an HPLC assay with a standard deviation of 1.3 for the main peak would result in a reporting interval of 0.1. Reporting intervals for impurities (minor peaks) need to be consistent with reporting intervals for the main peak. In general, STD of equal or less than 2 (in units reported by the method) is required to ensure a reporting interval of one decimal place. The argument can be raised that for low level, minor analytes (for example, the dimer in SEC present at 1%), the requirement for STD to be at or below 2% is too generous. This will result in an RSD of 200% for the peak. In such an instance, this would indicate that the minor peak is well below the detection level, because theoretically the RSD at the LOD level should not exceed 33% (Long and Winefordner 1983; Hayashi and Matsuda 1995)

The table below proposes the nearest reporting intervals based on standard deviations obtained during qualification for protein products.

Standard Deviation (in reported units)	Nearest Reporting Interval
≤2.0	0.1
≤20	1

Table 2. Recommended Nearest Reporting Results Based on Standard Deviation

Method precision is closely linked to the concentration of the analyte. The best-known relationship between analyte concentration and RSD is the Horwitz equation (Horwitz 1982; Horwitz and Albert 1997; Horwitz and Albert 1997)

$$RSD = 2^{(1-0.5 \log C)}$$

where, C is the concentration of the analyte in mg/g.

Based on the Horwitz equation, the precision of the measurement, expressed as RSD, doubles for each decrease of analyte concentration of two orders of magnitude.

The Horwitz relationship can provide good guidance for method precision targets during method development and qualification. Intermediate precision obtained during these studies should meet the variability derived from the Horwitz equation for each individual analyte. If, during execution of the qualification experiments, the precision of the measurements exceeds values derived from the Horwitz equation, this may indicate that the assay may need to be redeveloped, or that the technology utilized in the assay may not be fully suitable for the intended application.

Typically, proteins are available for analysis as solutions, with concentrations ranging widely from 1 µg/ml (e.g., a growth factor) to 100 mg/ml or higher (e.g., a monoclonal antibody). In such cases, expectations for the RSD of measurements of the main protein analyte in these solutions, based on the Horwitz relationship (e.g., using protein concentration method), will be 16 and 2.8 %, respectively.

As mentioned earlier, the precision of the method is referred to as uncertainty. The uncertainty of results is a parameter that describes a range within which the measured value is expected to lie (Miller and Miller 2000). Intuitively, we associate this parameter with precision. Therefore, method precision has been viewed as the most important performance characteristic. Typically, method precision has been assessed from replicate analyses of the same sample. The work of Hayashi and Matsuda on FUMAI theory (Hayashi, Rutan et al. 1993; Hayashi and Matsuda 1994; Hayashi and Matsuda 1994; Hayashi and Matsuda 1995; Hayashi, Matsuda et al. 2002; Hayashi, Matsuda et al. 2004) demonstrated that the precision of chromatographic methods can be predicted from noise and the height and width of the signal (peak). However, due to the complexity associated with the required Fourier transformation of chromatograms and the parameterization of the power spectrum called for in implementation of this theoretical construct to the determination of precision, the FUMAI theory approach has not been widely applied.

Apostol et al. (Apostol, Kelner et al. 2012) proposed a new approach to assessing the uncertainty of purity analyses that uses a more holistic approach that is called Uncertainty Based on Current Information (UBCI). The model allows for real-time assessment of all performance characteristics using the results of the specific separation of interest. A fundamental, underlying principle of this approach recognizes that the execution of a purity method is always associated with specific circumstances; therefore, uncertainty about the generated results needs to account for both the operational conditions of the method and the hardware. The authors demonstrated that noise levels, instrument and software settings can be linked directly to all method performance characteristics. Such simplification makes it easy to implement this procedure in a daily operation, and can provide a valuable live assessment of uncertainty instead of extrapolating uncertainty from historical qualification/validation studies.

The UBCI model approximates the maximal uncertainty of the measurement associated with the actual conditions of analysis (test). The obtained precision corresponds to the uncertainty under the most unfavorable conditions, including the highest variability of injection, maximal numeric integration error, expected variability of the peak width, and the most unfavorable contribution of the noise. UBCI shows that the uncertainty of results is not only a function of the method (composition of the mobile phase, gradient, flow rate, temperature), but also is influenced by the hardware associated with the execution of the method (pump pulsation, detector range, status of the lamp, etc.), and the software settings used to acquire the output in the form of chromatograms. Information about these parameters can be extracted from individual chromatograms; therefore, the assessment of method performance characteristics (uncertainty) can be performed real-time, which can be considered as a 'live validation' associated with each individual test result.

It is important to note that historical qualification/validation approaches do not take this fundamental principle into account, such that performance drift may occur over time due to hardware differences and even due to differences in analyst skill levels, such that the uncertainty of results obtained early in the product lifecycle may not be fully applicable to results obtained later. Application of historical validation data always begs a question about the relevance of these data to the current experimental situation, and sometimes requires investigation, which can delay the approval of results. The UBCI approach, therefore, has the capability of providing not only simplicity, but also a greater level of assessment of the data validity relative to current practices.

7. Accuracy

The determination of accuracy for protein purity methods presents significant challenges. Since it is difficult to establish orthogonal methods for proteins to measure the same quality attribute, it is hard to assess the truthfulness of the accuracy measurements. For example, although SEC-HPLC results can be verified by analytical ultra centrifugation (AUC) techniques, these techniques are based on very different first principles, and may not provide comparable results (Carpenter, Randolph et al. 2010; Svitel, Gabrielson et al. 2011). Therefore, in most cases, the accuracy of purity methods for proteins is inferred when other performance characteristics meet expectations, which is consistent with the principles of ICH Q2R1(ICH 2005).

8. Linearity and range

Linearity and range are typically assessed in a complex experiment demonstrating a linear change of peak area with analyte concentration. Since most of the methods use UV detection, such linearity experiments can be considered as re-confirmation of the Beer-Lambert law for the particular hardware configuration.

9. Specificity

The specificity of analytical methods is typically assessed by examining system interference with the detection and quantification of analytes. Part of this evaluation is the determination of protein recovery from the column (Rossi, Pacholec et al. 1986; Eberlein 1995). The recovery determination requires the knowledge of the extinction coefficient for the protein, which can be calculated from its amino acid composition (Pace, Vajdos et al. 1995) or determined experimentally. It should be noted that the extinction coefficient of a protein may change as a function of pH (Eberlein 1995; Kendrick, Chang et al. 1997). Therefore, direct comparison of the recovery in the neutral pH, size exclusion method with the recovery in an acidic reversed-phase separation may not be valid due to differences in the operating pHs of the methods. The difference may not necessarily reflect the actual recovery, but rather shows pH dependent changes of spectroscopic properties of the protein. With such an approach, the specificity of the method can be assessed in every assay, and reflects dynamically the change in status of consumables (columns and mobile phases) and hardware.

10. LOD and LOQ

Assessment of limit of detection/limit of quantitation (LOD/LOQ) is required for most analytical methods developed to monitor product quality attributes. In the latest version of ICH Q2R1, the terms Detection Limit (LD) and Quantitation Limit (QL) are used instead of LOD and LOQ, respectively. ICH defines LOD (DL) as the minimum level of analyte which can be readily detected, while LOQ (QL) has been defined as the minimum level of analyte which can be quantified with acceptable accuracy and precision. Practical application of LOD is related to the decision about integration of chromatograms, electropherograms or spectra, while LOQ is related to the decision on whether to report the results of tests on official documents, such as the Certificate of Analysis (CoA) for the lot.

ICH Q2R1 suggests three different approaches: visual inspection, signal-to-noise-ratio, or variability of the slope of the calibration curve (statistical approach). Vial and Jardy, and Apostol et al., evaluated different approaches for determining LOD/LOQ and concluded that they generate similar results (Vial and Jardy 1999; Apostol, Miller et al. 2009). It is prudent to verify LOD/LOQ values obtained by different calculations. If those values are not within the same order of magnitude, then the integrity of the source data should be investigated.

The statistical approach is most commonly practiced, and is associated with the use of well known equations:

$$LOD = 3.3 \times SD/S$$
$$LOQ = 10 \times SD/S$$

SD = standard deviation of response

S = slope of calibration curve (sensitivity)

The SD can be easily obtained from linear regression of the data used to create the calibration curves. The most common way to present calibration data for the purpose of linear regression is to graph the expected analyte concentration (spiked, blended) vs. the recorded response (UV, Fl, OD etc). This type of plot is characteristic of analytical methods for which the response is a linear function of the concentration (e.g. UV detection that follows the Beer-Lambert law). In cases where the measured response does not follow a linear dependency with respect to concentration (e.g., multi-parameter fit response of immunoassays), the response should be transformed to a linear format, such as semi-logarithmic plots, so that the equations above can be utilized.

The slope used in these equations is equivalent to instrument sensitivity for the specific analyte, reinforcing the fact that LOD/LOQ are expressed in units of analyte concentration (e.g. mg/ml) or amount (e.g., mg). Since the LOD and LOQ are functions of instrument sensitivity, these values, when defined this way, are not universal properties of the method transferable from instrument to instrument, or from analyte to analyte.

Considering LOD from the perspective of the decision to include or disregard a peak for integration purposes, stresses the importance of signal-to-noise ratio as a key parameter governing peak detection. Defining LOD as 3.3 x noise creates a detection limit, which can serve as a universal property of methods applicable to all analytes and different instruments (because sensitivity factor has been disregarded in this form of the equation). LOD expressed in this format is a dynamic property due to the dependency on the type of instrument, status of the instrument, and quality of the consumables. LOD determined this way will be expressed in units of peak height, e.g. mV or mAU.

The decision about reporting a specific analyte on the CoA is typically linked to specifications. After the decision about integration has been made for all analytes resolved (defined) by the method, the results are recorded in the database (e.g. LIMS). When all analytical tests are completed, the manufacturer creates the CoA by extracting the relevant information from the database. Only a subset of the results, which are defined by specifications, will be listed on the CoA. The specifications will depend on the extent of peak characterization and the clinical significance of the various peaks (Apostol, Schofield et al. 2008). Therefore, the list will change (evolve) with the stage of drug development. In such a context, LOQ should be considered as the analyte specific value expressed in units of protein concentration, a calculation for which instrument sensitivity cannot be disregarded (in contrast to LOD estimation). This indicates that a potential exists for diverse approaches to the practice of determining the LOD and LOQ.

Application of LOD/LOQ to purity methods presents specific challenges that deserve additional consideration. The reporting unit for purity methods is percent (%) purity, a unit that is not compatible with the unit in which LOD or LOQ are typically expressed (units of concentration or amount). The signal created by the analyte may vary with the load, while the relative percentage of the analyte does not change. This creates a situation where the analyte of interest can be hidden within the noise or, alternatively, can be significantly above the noise for the same sample analyzed at two different load levels within the range allowed by the method. This has been addressed by the concept of "dynamic LOQ" by combining statistical and S/N approaches (Apostol, Miller et al. 2009).

$$LOQ = 10 \times (S/N)^{-1} \times P$$

N = level of peak-to-peak noise
S = peak height for the analyte of interest
P = purity level for the analyte of interest

The above equation expresses LOQ as a function of signal-to-noise ratio and the observed purity of the analyte. Both parameters can change from test-to-test, due to equipment variability and sample purity variability. Therefore this equation should be viewed as the dynamic (live) assessment of LOQ.

11. System suitability

System suitability is intended to demonstrate that all constituents of the analytical system, including hardware, software, consumables, controls, and samples, are functioning as required to assure the integrity of the test results. System suitability testing is an integral part of any analytical method, as specified by ICH Q2R1. However, guidance is vague and reference is often made to Pharmacopeias for additional information. The USP, EP and JP contain guidance for a broad scope of HPLC assays, including assays of the active substance or related substances assays, assays quantified by standards (external or internal) or by normalization procedures, and quantitative or limit tests. While each type of assay is described in the compendia, the specific system suitability parameters to be applied for each type of assay, is not included with the description. Thus, some interpretation is required. The interpretation of how to best meet the requirements of the various compendia while still maintaining operational efficiency is a significant challenge for industry.

Existing guidance for system suitability was developed for pharmaceutical compounds and may not be directly applicable for proteins which, due to their structural complexity and inherent heterogeneity, require additional considerations beyond those typically required for small molecules. For example, appraisal of resolution by measuring the number of theoretical plates (commonly done for small molecules), may not be the best way to assess the system readiness to resolve charge isoforms of a protein on an ion exchange column. This may be due to the relatively poor resolution of protein peaks resulting from inherent product microheterogeneity, when compared to the resolution typically seen with small molecules. However, this methodology (the number of theoretical plates) may be a very good indicator to measure the system performance for size exclusion chromatography (SEC), which does not typically resolve product isoforms resulting from microheterogeneity.

To appropriately establish system suitability, we need to consider both the parameter that will be assessed and the numerical or logical value(s), generally articulated as acceptance criteria, associated with each parameter. System suitability parameters are the operating parameters that are the critical identifiers of an analytical method's performance. System suitability should be demonstrated throughout an assay by the analysis of appropriate controls at appropriate intervals. It is a good practice to establish the system suitability parameters during method development, and to demonstrate during qualification that these parameters adequately evaluate the operational readiness of the system with regard to such factors as resolution, reproducibility, calibration and overall assay performance. Prior to validation, the system suitability parameters and acceptance criteria should be reviewed in order to verify that the previously selected parameters are still meaningful, and to establish limits of those parameters, such that meaningful system suitability for validation is firmly established.

One important issue that merits consideration is that the setting of appropriate system suitability parameters is a major contribution to operational performance in a Quality environment, as measured by metrics such as invalid assay rates. A key concept is that the purpose of system suitability is to ensure appropriate system performance (including standards and controls), not to try to differentiate individual sample results from historical trends (e.g., determining equivalence of results from run-to-run). In practice, setting system suitability parameters that are inappropriately stringent can result in the rejection of assay results with acceptable precision and accuracy. It is highly advisable to ensure the participation of Quality Engineers and/or other staff members with appropriate statistical expertise when setting system suitability parameters.

12. Method robustness

ICH Q2R1 prescribes that the evaluation of robustness should be considered during the development phase. The robustness studies should demonstrate that the output of an analytical procedure is unaffected by small but deliberate variations in method parameters. Robustness studies are key elements of the analytical method progression and are connected to the corresponding qualification studies.

Method robustness experiments cannot start before the final conditions of the method are established. It is a good practice to identify operational parameters for the method and to divide them in the order of importance into subcategories according to their relative importance, which are exemplified below:

1. "Essential" category: includes method parameters that are critical to the method output and therefore require evaluation;
2. "Less important" category: includes method parameters that are not as critical as those in the "Essential" category, but may still affect the method output. These parameters should be evaluated at the scientist's discretion;
3. "Depends on Method" category: includes parameters that may affect the method output differently for different methods, such that these parameters should be treated differently for each method;
4. "Not useful" category: includes method parameters that are known to have no impact on the method output, such that these parameters need not be evaluated for robustness.

It is highly impractical to evaluate the impact of all possible parameters on the output of the method. Therefore, robustness studies could be limited to the demonstration that the reported assay values are not affected by small variations of "essential" operational parameters. It is a good practice to prospectively establish a general design (outline) for such studies. Typically, in these types of studies a reference standard and/or other appropriate samples are analyzed at the nominal load. The studies may be carried out using the one-factor-at-a-time approach or a Design of Experiment (DOE) approach. The selection of assay parameters can vary according to the method type and capabilities of the factorial design, if applicable. It is essential to study the impact of all essential factors, and it is important to establish prospectively "target expectations" for acceptable changes in the output, to ensure that these robustness studies do not repeat the development work. The maximum allowable change in the output of the analytical method can be linked to the target expectations for the precision of the method, which are derived from the Horwitz equation (Horwitz 1982; Horwitz and Albert 1997; Horwitz and Albert 1997). Recently a number of software packages have become available to assist with the design and data analysis (Turpin, Lukulay et al. 2009; Jones and Sall 2011; Karmarkar, Garber et al. 2011).

13. Challenges associated with validated methods

Remediation of validated analytical methods is typically triggered by the need to improve existing methods used for disposition of commercial products. The improvement may be required due to an unacceptable rate of method failures in the GMP environment, lengthy run times, obsolete instruments or consumables, the changing regulatory environment for specifications or stability testing, or for other business reasons.

We anticipate that technological advances will continue to drive analytical methods toward increasing throughput. In this context, it appears that many release methods are destined for change as soon as the product has been approved for commercial use (Apostol and Kelner 2008; Apostol and Kelner 2008). This is due to the fact that it takes more than 10 years to commercialize a biotechnology drug, resulting in significant aging of the methods developed at the conception of the project. Therefore, the industry and regulators will need to continuously adjust strategies to address the issue of old vs. new methods, particularly with respect to how these advances impact product specifications (Apostol, Schofield et al. 2008). Frequently, old methods have to be replaced by methods using newer technologies, creating a significant challenge for the industry in providing demonstration of method equivalency and a corresponding level of validation for the methods.

14. Concluding remarks

When we consider the critical role that analytical method development, qualification and validation play in the biopharmaceutical industry, the importance of a well designed strategy for the myriad analytical activities involved in the development and commercial production of biotechnology products becomes evident.

The method qualification activities provide a strong scientific foundation during which the performance characteristics of the method can be assessed relative to pre-established target expectations. This strong scientific foundation is key to long-term high performance in a Quality environment, following the method validation, which serves as a critical pivotal point in the product development lifecycle. As noted previously, the method validation often serves as the point at which the Quality organization assumes full ownership of analytical activities. If done properly, these activities contribute to operational excellence, as evidenced by low method failure rates, a key expectation that must be met to guarantee organizational success. Without the strong scientific foundation provided by successful method development and qualification, it is unlikely that operational excellence in the

Quality environment can be achieved. As analytical technologies continue to evolve, both the biotechnology industry and the regulatory authorities will need to continuously develop concepts and strategies to address how new technologies impact the way in which the Quality by Design principles inherent in the analytical lifecycle approach are applied to the development of biopharmaceutical products. The basic concepts are described in ICH guidelines Q8, Q9 and Q10 (ICH 2005; ICH 2008).

The ICH Q2 guideline requires that an analytical method be validated for commercial pharmaceutical and bio-pharmaceutical applications. Frequently, validation is done only once in the method's lifetime. This is particularly of concern when the future testing is performed on an instrument with different technical characteristics, in different geographic locations within the company and/or at contract laboratories around the world, using different consumables, different analysts, etc. This concern is exacerbated by the requirement for modern pharmaceutical and biopharmaceutical companies to seek regulatory approval in multiple jurisdictions, where the instrumentation, consumables, and scientific staff experience at the testing location may be very different than that present in the place where the drug was developed. These considerations raise questions about the value of the current format of the validation studies conducted by the industry. Moreover, it is not clear how the validation data obtained using existing methodologies should or even could be used toward the assessment of the uncertainty of the future results, given the many factors that contribute to the uncertainty.

Perhaps the time is right for the industry to consider the use of a combination of sound science and reasonable risk assessment to change the current practice of the retrospective use of method validation to the new paradigm of live validation of purity methods based on the current information embedded in the chromatogram. Laboratories that work in a GMP environment are required to produce extensive documentation to show that the methods are suitable. Pharmaceutical and biopharmaceutical companies thoroughly adhere to these requirements, inundating industry with an avalanche of validation work that has questionable value toward the future assessment of uncertainty. The predication of uncertainty provides an alternative that has the potential to reduce the work required to demonstrate method suitability and, in turn, provide greater assurance of the validity of the results from the specific analysis in real time.

The establishment of qualification target expectations can be considered as a form of Quality by Design (QbD), since this methodology establishes quality expectations for the method in advance of the completion of method development. Also, the analytical lifecycle described here covers all aspects of method progression, starting with method development, the establishment of system suitability parameters, and qualification and robustness activities, culminating in method validation, which confirms that the method is of suitable quality for testing in Quality laboratories. The entire analytical lifecycle framework can be considered as a QbD process, consistent with evolving regulatory expectations for pharmaceutical and biopharmaceutical process and product development.

Author details

Izydor Apostol, Ira Krull and Drew Kelner
Northeastern University, Boston, MA, USA

15. References

Agut, C., A. Segalini, et al. (2006). "Relationship between HPLC precision and number of significant figures when reporting impurities and when setting specifications." J Pharm and Biomed Analysis 41(2): 442-448.

Apostol, I. and D. Kelner (2008). "Managing the Analytical Lifecycle for Biotechnology Products: A journey from method development to validation, Part 1." BioProcess International 6(8): 12-19.

Apostol, I. and D. Kelner (2008). "Managing the Analytical Lifecycle for Biotechnology Products: A journey from method development to validation, Part 2." BioProcess International 6(9): 12-19.

Apostol, I., D. Kelner, et al. (2012). "Uncertainty estimates of purity measurements based on current information: toward a "live validation" of purity methods." Pharmaceutical Research in press.

Apostol, I., K. J. Miller, et al. (2009). "Comparison of different approaches for evaluation of the detection and quantitation limits of a purity method: a case study using a capillary isoelectrofocusing method for a monoclonal antibody." Anal Biochem 385(1): 101-106.

Apostol, I., T. Schofield, et al. (2008). "A Rational Approach for Setting and Maintaining Specifications for Biological and Biotechnology Derived Products - Part 1." BioPharm International 21(6): 42-54.

ASTM (2005). "Standard Practice for Using Significant digits in Test Data to Determine Conformance with Specifications." ASTM International In ASTM Designation: E 29-04: 1-5.

Carpenter, J. F., T. W. Randolph, et al. (2010). "Potential inaccurate quantitation and sizing of protein aggregates by size exclusion chromatography: essential need to use orthogonal methods to assure the quality of therapeutic protein products." J Pharm Sci 99(5): 2200-2208.

CDER (2001). "*Guidance for industry Bioanlaytical method validation.*" *Guidance for industry: Bioanlaytical method validation.*

Eberlein, G. A. (1995). "Quantitation of proteins using HPLC-detector response rather than standard curve comparison." Journal of Pharmaceutical and Biomedical Analysis 13(10): 1263-1271.

Hayashi, Y. and R. Matsuda (1994). "Deductive prediction of measurement precision from signal and noise in liquid chromatography." Anal. Chem. 66: 2874-2881.

Hayashi, Y. and R. Matsuda (1994). "Deductive prediction of measurement precision from signal and noise in photomultiplier and photodiode detector of liquid chromatography." Analytical Sciences 10: 725-730.

Hayashi, Y. and R. Matsuda (1995). "Prediction of Precision from Signal and Noise Measurement in Liquid Chromatography - Mathematical Relationship between Integration Domain and Precision." Chromatographia 41(1-2): 75-83.

Hayashi, Y., R. Matsuda, et al. (2002). "Validation of HPLC and GC-MS systems for bisphenol-A leached from hemodialyzers on the basis of FUMI theory." J Pharm Biomed Anal 28(3-4): 421-429.

Hayashi, Y., R. Matsuda, et al. (2004). "Precision, Limit of Detection and Range of Quantitation in Competitive ELISA." Analytical Chemistry 76(5): 1295-1301.

Hayashi, Y., S. C. Rutan, et al. (1993). "Information-Based Prediction of the Precision and Evaluation of the Accuracy of the Results from an Adaptive Filter." Chemometrics & Intelligent Laboratory Systems 20(2): 163-171.

Holme, D. M. and H. Peck (1998). Analytical Biochemistry. Upper Saddle River, New Jersey, Prentice-Hall, Inc,. Third Edition: 9-19.

Horwitz, W. (1982). "Evaluation of Analytical Methods Used for Regulation of Foods and Drugs." Analytical Chemistry 54: 67-76.

Horwitz, W. and R. Albert (1997). "A Heuristic Derivation of the Horwitz Curve." Analytical Chemistry 69: 789-790.

Horwitz, W. and R. Albert (1997). "Quality IssuesThe Concept of Uncertainty as Applied to Chemical Measurements." Analyst 122(6): 615-617.

ICH (2005). "ICH Q2R1, Validation of Analytical Procedures." Harmonised Tripartite Guideline: 1-13.

ICH (2005). "ICH Q9, Quality Risk Management." ICH Harmonised Tripartite Guideline.

ICH (2006). Analytical Methods Validation, Advanstar Communications.

ICH (2008). "ICH Q10, Pharmaceutical Quality System." ICH Harmonised Tripartite Guideline.

Jiang, Y., C. Li, et al. (2012). Qualification fo biophysical methods for the analysis of protein therapeutics. Biophysics for Therapeutic Protein Development. L. Narhi, Springer.

Jones, B. and J. Sall (2011). "JMP statistical discovery software." Wiley Interdisciplinary Reviews: Computational Statistics 3(3): 188-194.

Karmarkar, S., R. Garber, et al. (2011). "Quality by Design (QbD) Based Development of a Stability Indicating HPLC Method for Drug and Impurities." Journal of Chromatographic Science 49(6): 439-446.

Kendrick, B. S., B. S. Chang, et al. (1997). "Preferential exclusion of sucrose from recombinant interleukin-1 receptor antagonist: role in restricted conformational mobility and compaction of native state." Proc Natl Acad Sci U S A 94(22): 11917-11922.

Krull, I. S. and M. Swartz (2006). "Validation in Biotechnology and Well-characterized Biopharmaceutical Products " Pharmaceutical Regulatory Guidance Book(July): 18-23.

Long, G. L. and J. D. Winefordner (1983). "Limit of Detection." Analytical Chemistry 55: 712A-724A.

Miller, J. N. and J. C. Miller (2000). Statistics and Chemometrics for Analytical Chemistry. Upper Saddle River, New Jersey, Prentice-Hall, Inc.

Pace, C. N., F. Vajdos, et al. (1995). "How to measure and predict the molar absorption coefficient of a protein." Protein Sci 4(11): 2411-2423.

Ritter, N. M., S. Advant, J., et al. (2004). "What Is Test Method Qualification? , Proceedings of the WCBP CMC Strategy Forum, 24 July 2003 ." BioProcess Int 2: 32-46.

Rossi, D. T., F. Pacholec, et al. (1986). "Integral method for evaluating component contribution to total solution absorbance from chromatogrphic Data." Anal. Chem. 58: 1410-1414.

Shabir, G. A. (2003). "Validation of high-performance liquid chromatography methods for pharmaceutical analysis: Understanding the differences and similarities between validation requirements of the US Food and Drug Administration, the US Pharmacopeia and the International Conference on Harmonization." Journal of Chromatography A 987(1-2): 57-66.

Svitel, J., J. P. Gabrielson, et al. (2011). "Analysis of self association of proteins by analytical ultracentrifugation." In prepration.

Swartz, M. E. and I. S. Krull (1997). *Analytical Method Development and Validation* New York, Marcel Dekker

Swartz, M. E. and I. S. Krull (2006). "Method Validation and Robustness." LCGC North America 24(5): 480-490.

Swartz, M. E. and I. S. Krull (2009). "Analytical Method Validation: Back to Basics, Part 1 " LCGC North America 27(11): 11-15.

Turpin, J., P. H. Lukulay, et al. (2009). "A Quality-by-Design Methodology for Rapid LC Method Development, Part III." LC-GC North America 27(4): 328-339.

USP (1994). Chromatography Section. United States Pharmacopeia Convention. Rockville, MD, US Pharmacopeia 1776.

USP (1999). Validation of Compendial Methods. Rockville, MD, US Pharmacopeia: 2149.

Vial, J. and A. Jardy (1999). "Experimental Comparison of the different approaches to estimate LOD and LOQ of an HPLC Method." Anal.Chem., 71: 2672-2677.

Chapter 6

Peptide and Amino Acids Separation and Identification from Natural Products

Ion Neda, Paulina Vlazan, Raluca Oana Pop,
Paula Sfarloaga, Ioan Grozescu and Adina-Elena Segneanu

Additional information is available at the end of the chapter

http://dx.doi.org/10.5772/51619

1. Introduction

The natural extracts of plants are an important source for the identification of new biologically active compounds with possible applications in the pharmaceutical field. Phytotherapy embraces especially the isolation from herbs, of compounds with unique chemical structures, which are considered to be pharmacologically active. Recent statistics show that, on an annual basis, there are identified over 1500 new compounds from different species of plants, and about one quarter of prescription drugs contain substances of plant origin.

Although globally there are many studies on natural products, research in this area continues to be of great latent potential, in part due to the existing and still unexploited, potential and perspective development of new opportunities to recovery in major industrial areas and because of their socio-economic impact. A large number of plant extracts that have been used in the traditional medicine have also applications nowadays, both in the pharmaceutical and food industries (Busquet et al., 2005 [1]).

It is known that native Carpathian flora represents about 30% of plant species on the entire European continent. In Romanian, traditional medicine, there are used a number of plants with powerful therapeutic effects, and, as a result, an extensive investigation of their content is necessary. There are many classes of compounds that can be found in an alcoholic, natural extracts: amino acids, peptides, small proteins, phenols, polyphenols, saponins, flavonoids and sugars. Compounds of great interest are the free amino acids and the peptides from these extracts, which show important antitumor activities.

The process of structure elucidation of a natural product involves the determination of many physical-chemical properties: melting point, optical rotation, solubility, absorption, optical rotatory dispersion, circular dichroism, infrared spectroscopy, as well as mass and nuclear magnetic resonance spectroscopies. On the basis of such information will be proposed a likely and reasonable structure(s) for the studied natural product.

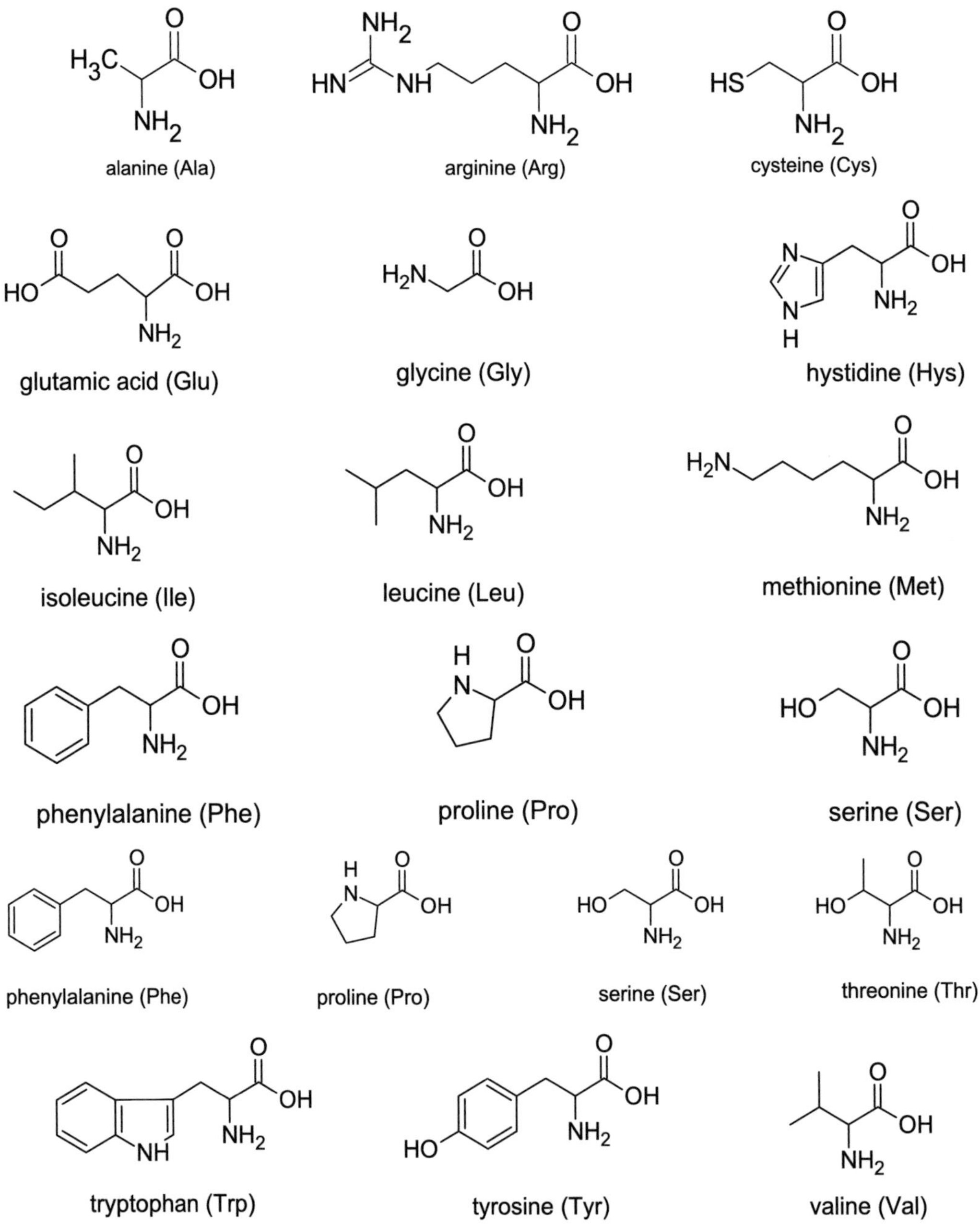

Figure 1. Amino acids chemical structures

If early methods for organic molecule characterisation used only a few physico-chemical parameters, such as: melting point, solubility, elemental analysis, molecular weight, and/or specific rotation, yet now more modern techniques, especially various spectroscopies, of analysis and characterization are extremely useful tools for a complete chemical screening of the natural extracts.

The chemistry of natural products includes three main areas: isolation, structure elucidation, and synthetic methods. The isolation step is considered to be a part of structure elucidation, and therefore analysis and characterization methods, such as the UV-Vis and infrared spectroscopy, mass spectrometry and various chromatographic techniques, are all important tools for a proper identification of the components of an extract. Figure 1 illustrates the chemical structures for most, widely found, naturally occurring amino acids.

2. Sample preparation

A preliminary step, required for the proper separation of amino acids and peptides, consists in finding a suitable, partitioning scheme of the extract between various solvents, in order to remove the unwanted compounds, such as: polysaccharides, lipids, phenols and others.

Capillary electrophoresis (CE) allows the separation of amino acids without prior derivatization. A derivatization step is often necessary in order to improve the detectability using optical detection. A wide variety of labeling reagents have been reported, such as: FMOC, NDA, OPA or FITC (fluorescein isothiocyanate).

Typically, in amino acid analysis, peptide bonds must first be broken, into the individual amino acid constituents. It is known, that the sequence and nature of amino acids in a protein or peptide determines the properties of the molecule. There are different hydrolyzing methods commonly utilized before amino acid analysis, but the most common is acid hydrolysis. However, some of the amino acids can be destroyed using such an approach. Thus, methionine and cystine were either partially destroyed or oxidised to methionine sulphone and cysteic acid. Usually, it is often best to use a hot hydrochloric acid solution and 0.1% to 1.0% of phenol, which is added to prevent halogenation of tyrosine.

Alkaline hydrolysis method has limited applications due the destruction of arginine, serine, threonine, cysteine and cystine. Enzymatic hydrolysis represents perhaps the best method for the complete hydrolysis of peptide bonds, because it does not affect tryptophan, glutamine and asparagines. However, their applications are restricted, due to the difficulties often involved with the use of enzymes.

Separation and elucidation of the chemical composition of a natural product, from a medicinal plant, involves a very laborious procedure. For instance, in the case of *Chelidonium majus L,* a well –known herb, it was necessary to perform successive extractions with hexane, ethyl acetate, chloroform, and n-butyl alcohol. Every fraction obtained was analyzed in detail by various spectroscopic and chromatographic techniques.

Recent scientific research has reported a number of increased and improved techniques for the identification of free amino acids, such as, spectroscopic identification by means of colorimetric methods. These have often used reagents such as 2,4-dinitrofluorobenzene [2] and genipin [3]. Also, it has also been reported on the use of IR spectroscopy for the study of various extracts of *Angelica* [4].

3. UV-Vis spectroscopy

The UV-Vis spectra of natural compounds contain information about different properties (such as: chemical composition and structure). Such methods are simple, fast, inexpensive, and safe to perform; which accounts for their popularity. However, these methods have disadvantages, because the result's accuracy depends on many factors: e.g., variations in the length of the polypeptide chain, amount and types of amino acid residues, accessibility of dye reagents, presence of final buffers, stabilizers, and other excipients, which can react with dyes or absorb at the detection wavelength.

4. IR spectroscopy

Infrared spectroscopy is based on molecular vibrations, characteristic to the specific chemical bonds or groups. The energy of most molecular vibrations (stretching, twisting and rotating) corresponds to that of the infrared region of the electromagnetic spectrum. There are many vibrational modes that do not represent a single type of bond oscillation but are strongly dependent on the neighbouring bonds and functional groups. One of the great advantages of this analytical technique for natural products, is due to the fact that spectra can be obtained form almost any environment (aqueous solution, organic solvents, etc.) and from relatively small quantities of sample.

There are a large number of IR spectroscopic studies regarding the structure of amino acids and peptides; some of the approached subjects are the following: infrared spectra of potassium ion tagged amino acids and peptides (Polfer et al., 2005[5]), IR spectra of deprotonated amino acids (Oomens et al., 2009[6]). Also, there have been reported studies regarding the IR spectra of some derivatives of the amino acids, namely the amides (Kasai et al., 1979[7]). The results show the appearance of the C=O group around 1675-1680 cm-1 for most of the studied compounds. An exception is represented by the L-tyrozine amide, which shows a vibration frequency of the C=O group at 1705 cm-1, fact that is probably due to the intermolecular hydrogen bonds between the N atom of the amidic group and the phenolic OH group (Kasai et al, 1979[7]). Linder et al. have presented the IR spectra of 5 natural amino acids, namely valine, proline, isoleucine, phenylalanine and leucine (Linder et al., 2005[8]). The five spectra are very similar as regards the position of C=O group, and the absorption frequencies of OH groups. Slight differences appear only in the case of the stretching vibrations of the C-H groups (Linder et al., 2005[8]).

The amino acid and peptide absorption bands in the 3400 cm^{-1} region is due to O–H and N–H, bond stretching. The broad absorption bands in the region 3030-3130 cm^{-1} are attributed to asymmetric valence vibrations of the ammonium (NH_3^{+}) group. The symmetric absorption vibrations in 2080 -2140 cm^{-1} or 2530-2760 cm^{-1}, depend on amino acid chemical structures. The ammonium group deformation vibrations are located at 1500-1600 cm^{-1}, together with the absorptions characteristic of the carboxylate ion. The asymmetrical deformation bands from 1610-1660 cm^{-1} is associated with a carboxylate (COO^-) group, and it usually represents a weak absorption. The bands in the 1724-1754 cm^{-1} region correspond to the carbonyl (C=O) vibration.

In the next figure (Figure 2), is presented the FT-IR spectra of L-leucine.

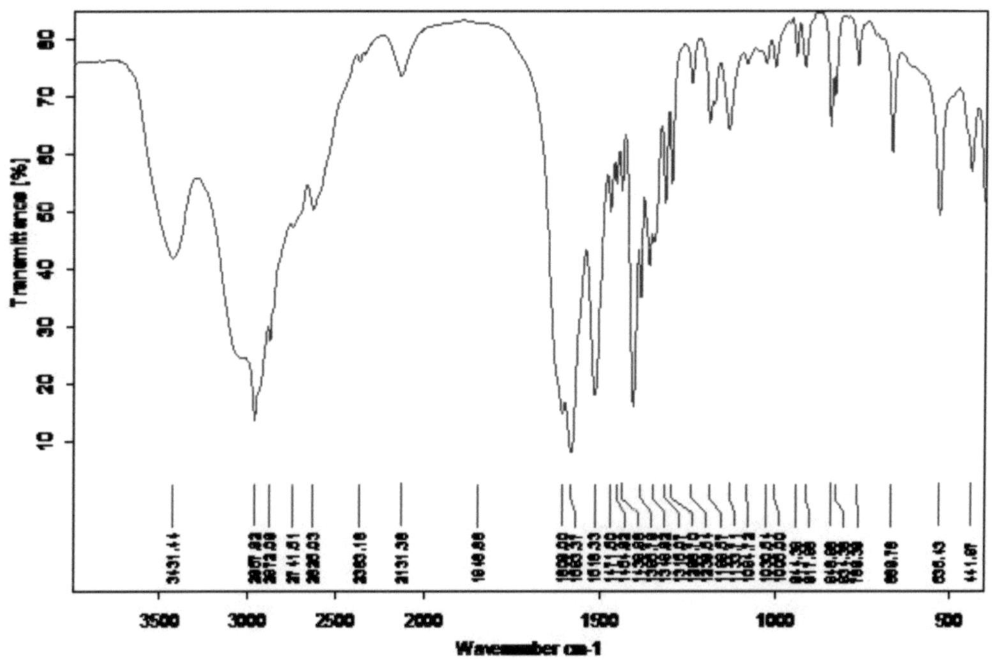

Figure 2. FT-IR spectra of leucine

In the following Figure 3, the IR spectrum of the *Chelidonium majus L.* extract is presented:

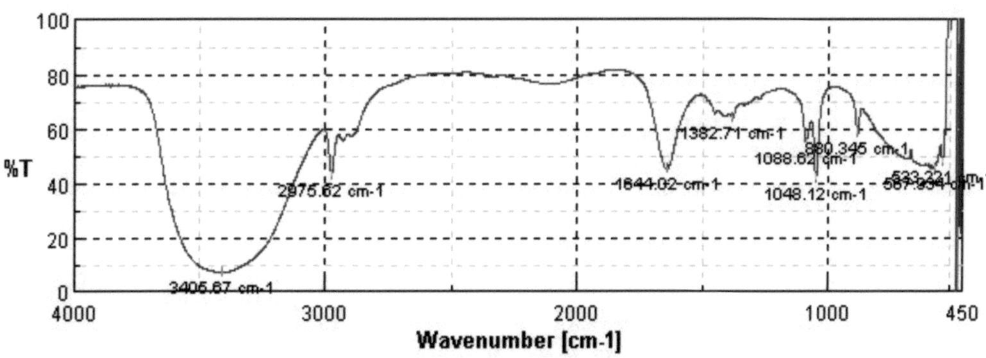

Figure 3. IR spectrum of the aqueous part of the *Chelidonium majus L.* extract (after successive extractions with hexane, ethyl acetate, chloroform and n-butyl alcohol)

The wavenumbers that appear in the IR spectra can be attributed to: OH (3405.67 cm^{-1}), CH_2 and CH_3 (2975.62 cm^{-1}), C=C (1644.02 cm^{-1}), and C-O (1382.71 cm^{-1}). Also, the UV-Vis spectra of the aqueous part of *Chelidonium majus L.*, showed the existence of three absorption bands: 734 nm, 268 nm and 198 nm, respectively. For a complete study, further analysis (including derivatization and HPLC) are usually performed.

5. Chromatographic methods

These techniques insure the separation of closely related compounds in a mixture, by differences in the equilibrium or partition distributions of the components between two immiscible phases, the stationary and the mobile phases. These differences in the equilibrium distribution are a result of chemical structures and the degree of interactions of the components between these two phases. Under the influence of a mobile phase (one or a mixture of solvents), the target compounds percolate through the stationary phase, which is a porous medium (usually, silica or alumina). For successful amino acid and peptide isolations and purifications from natural products, have been developed different chromatographic methods (e.g., paper, thin layer, gas chromatography, column and high performance liquid chromatography, etc.). From the enormous variety of methods of separation and isolation useful for natural products, adsorption or partition chromatography represents one of the most useful techniques of general application.

Thin layer chromatographic (TLC) is the simplest technique used to separate and identify natural products of interest. This method readily provides qualitative information and possibly quantitative data. The stationary phase is usually silica gel on the TLC or HPTLC (high performance TLC) plate, which is made up of silica adhered to glass or aluminium or a plastic, for support. The eluent (solvent mixture) acts as the mobile phase. Practically, the compounds of interest need to be soluble to varying degrees. Separation again results from the partition equilibrium of the components in the mixture. The separation depends on several factors: 1) solubility in the mobile phase, 2) attractions or adsorption between the compound and silica, the more the compound interacts with silica, the less it moves upwards, 3) size or MW of the compound, for the larger the compound, the slower it moves up the plate.

Since amino acids are colourless compounds, ninhydrin is routinely used to detect them, with the result of a coloured product, due to the formation of Ruhemann'purple complex. The familiar violet color which is associated with the reaction of amino acids with ninhydrin is attributed to the anion of the reagent (derivatizing agent). Another technique uses anisaldehyde-H2SO4 reagent for detection of amino acids, followed by heating (120° C, 5 minutes).

Different organic solvents (e.g., alcohol, dioxane, methyl cellosolve, pyridine, and phenol) are used to accelerate the development of color, to varying degrees. Ultimately a phenol-pyridine system was adopted as the most effective solvent. Exposure to 105°C for 3 to 5 minutes gives quantitative yields of color for all amino acids, except for tryptophan and lysine. The Rf (retardation factor) value for each compound can be calculated and compared with their reference values, in order to identify specific amino acids. The Rf value for each known compound should remain the same, provided the development of the plate is done with the same solvents, type of TLC plates, method of spotting and under exactly the same conditions.

High performance liquid chromatography (HPLC) allows for the most efficient and appropiate separations of consitutents from natural product, complex mixtures. It has been shown that HPLC is the premier separation method that can be used for amino acid analysis (AAA), from natural products, allowing for the separation and detection by UV absorbance or fluorescence. However, most common amino acids do not contain a chromophoric group, and thus some form of derivatization is usually required before HPLC or post-column.

Amino acids are highly polar molecules, and therefore, conventional chromatographic methods of analysis, such as, reversed-phase high performance liquid chromatography (RP-HPLC) or gas-chromatography (GC) cannot be used without derivatization. The derivatization procedure has several goals, such as: to increase the volatily, to reduce the reactivity, or to improve the chromatographic behaviour and performance of compounds of interest. In the case of amino acids, derivatization replaces active hydrogens on hydroxyl, amino and SH polar functional groups, with a nonpolar moiety. The great majority of derivatization procedures involve reaction with amino groups: usually primary amines, but also secondary amines (proline and hydroxyproline), or the derivatization of a carboxyl function of the amino acids. Some of the most common derivatization reagents are presented in the Table 1.

As it was mentioned before, prior derivatization of the amino acids is necessary due to the lack of UV absorbance in the 220-254 range. The paper of Moore and Stein [9] is actual even nowadays. Their method, that used a modified nynhidrin reagent for the photometric determination of the amino acids, represents the basis for various derivatization methods. There is a continuous increasing number of amino acids derivation reagents. There will be mentioned, as follows, some of the them: Melucci et al. [10] presents a method for the quantization of free amino acids that implies a pre-column derivatization with 9-fluorenylmethylchloroformate, followed by separation by reversed-phase high-performance liquid chromatography. Kochhar et al. [11] use the reverse-phase high-performance liquid chromatography for quantitative amino acids analysis and, as derivatization agent, 1-fluoro-2,4-dinitrophenyl-5-L-alanine amide, known as Marfey's reagent. The method was successfully applied for the quantization of 19 L-amino acids and it is based on the stoichiometric reaction between the reagent and the amino group of the amino acids [11]. Ngo Bum et al. [12] have been used the cation exchange chromatography and post-column derivatization with ninhydrin for the detection of the free amino acids from the plant extracts. Culea et al. [13] have used the derivatization of amino acids with trifluoroacetic anhydride, followed by the extraction with ion exchangers and GC/MS analysis. Warren proposed a version of CE, CE-LIF, for quantifying the amino acids from soil extracts. The advantage of the method is represented by the low detection limits that are similar to the ones corresponding to the chromatographic techniques. In 2010, Sun et al. [15] have presented another method for the detection of amino acids from Stellera chamaejasme L., a widely-used plant in the Chinese traditional medicine; DBCEC (2-[2-(dibenzocarbazole)-ethoxy] ethyl chloro-formate) was used as derivatization reagent, and the modified amino acids were detected by means of liquid chromatography with fluorescence detection. Li et al. [16] have proposed a new method for the detection of amino acids from the asparagus tin. After performing the derivatization of the samples with 4-chloro-3,5-dinitro-benzotrifluoride (CNBF), solid phase extractions on C18 cartridges have been performed. The purified amino acid derivatives were then subjected to the HPLC analysis. Zhang et al.[17] have proposed an improved chromatographic method (by the optimization of mobile phases and gradients) for the simultaneous detection of 21 free amino acids in tea leaves.

DERIVATIZATION REAGENT	Reference
Nynhidrin	9, 12
9-fluorenylmethylchloroformate	10
trifluoroacetic anhydride	13
1-fluoro-2,4-dinitrophenyl-5-L-alanine amide	11
2-[2-(dibenzocarbazole)-ethoxy] ethyl chloroformate	15
4-chloro-3,5-dinitrobenzotrifluoride	16
ortho-phthaldehyde (OPA)	
phenylisothiocyanate	
dimethylamino-azobenzenesulfonyl chloride	
7-fluoro-4-nitro-2-aza-1,3-diazole	

Table 1. Reagents for Derivatization

There have been already developed, several liquid chromatography methods for amino acid quantification. General approaches are ion-exchange chromatography (IEC) and reversed-phase HPLC (RP-HPLC). Both approaches require either a post-column or pre-column derivatization step. Even this technique offers satisfactory resolutions and sensitivity, but the necessary derivatization step provides an increased complexity, cost, and analysis times.

Ion-exchange chromatography with postcolumn ninhydrin detection is one of the most commonly used methods employed for quantitative amino acid analysis. Separation of the amino acids on an ion-exchange column is accomplished through a combination of changes in pH and ionic (cation) strength. A temperature gradient is often employed to enhance separation.

But, perhaps the most effective method is cation exchange chromatography (CEC) in the presence of a buffer system (usually a lithium buffer system), and a post-column derivatization step with ninhydrin. Detection is performed with UV absorbance. In this way one achieves the desired amino acid separation, according to the colour (structure) of the derivatived compound formed. Amino acids which contain primary amines, except an imino acid, give a purple color, and show the maximum absorption at 570 nm. The imino acids such as proline give a yellow color, and show the maximum absorption at 440 nm. The postcolumn reaction between ninhydrin and an amino acid eluted from the column is monitored at 440 and 570 nm.

OPA is another reagent used both for post-column or pre-column derivatization. Ortho-phthaldehyde (OPA) reacts at an amino group, generally in the presence of a thiol (mercaptoethanol), resulting in a fluorescent derivative, UV active at 340 nm. Other reagents for the precolumn derivatization of free amino groups from amino acids, are: PITC (phenylisothiocyanate), DABS-Cl (dimethylamino-azobenzenesulfonyl chloride), Fmoc-Cl (9-fluorenylmethyl-chloroformate), NBD-F (7-fluoro-4-nitro-2-aza-1,3-diazole), and others. The reaction time depends on the type of derivatization reagent and the reacting, functional group involved. For instance, from practically nearly instantaneous derivative formation in the case of the reagent, fluorene chloroformate, OPA is 1 minute and PITC is about 20 minutes.

Ion pair, reverse phase liquid chromatography coupled with mass spectroscopy, IPRPLC-MS/MS, is a technique which allows for the analysis of amino acids without derivatization, thus reducing the possible errors introduced by reagent, interferences and derivative instability, side reactions, etc. Using volatile reagents, the IP separation is based on two different mechanisms: a) the IP-reagent is adsorbed at the interface between stationary and mobile phases; and b) the formation of a diffuse layer and the electrostatic surface potential depends on superficial (surface) concentration of IP reagent. There are other, possible mechanisms suggested in the literature for how IPRPLC operates.

Gas chromatography (GC) can be used for the separation and analysis of compounds that can be vaporized without decomposition. The derivatization procedure most commonly employed is silylation, a method through which acidic hydrogens are replaced by an alkylsilyl group. Typically, silylation reagents are: BSTFA (*N,O*-bis-(tri-methyl-silyl)-trifluoroacetamide and MSTFA (*N*-methyl-silyl-trifluoro-acetamide). A possible disadvantage of this approach, is due to the reagent and derivative being sensitive to moisture and possible, derivative instability. Some amino acids are unstable (e.g., arginine and glutamic acid). Arginine descomposes to ornithine, and glutamic acid undergo a rearrangement to pyro-glutamic acid. Another GC-derivatization method includes acylation or esterification, now using an aldehyde and alcohol (pentafluorpropyl or trifluoracetic aldehyde and isopropanol) or alkyl chloroformate and alcohol. Silylation takes place through the direct conversion of carboxylic groups to esters and amino groups to carbamates. Such reactions are catalyzed by a base (pyridine or picoline). Alkyl esters are extremely stable and can be stored for long periods of time.

GC-MS represents an analysis method with excellent reproducibility of retention times, and the method can be easily automated. The major disadvantage is due to the possible temperature instabiity of some compounds and/or their derivatives, which then cannot be easily analyzed under most GC conditions.

Mass spectrometry represents one of most efficient techniques for natural product, structure elucidation. It functions by a separation of the ions formed in the ionisation source of the mass spectrometer, according to their mass-to-charge (m/z) ratios. The technique allows for accurate MW measurements, sample confirmation, demonstration of the purity of a sample, verification of amino acid substitutions, and amino acid sequencing. This procedure is useful for the structural elucidation of organic compounds and for peptide or oligo-nucleotide sequencing. The major advantadge in using MS is due to the need for very small amounts of sample (ng to pg). A disadvantage of conventional ionization methods (e.g., electron impact, API) is that they are limited to compounds with sufficient volatility, polarity and MW. Volatility can be increased by chemical modifications (derivatizations, such as: methylation, trimethylsilylation or trifluoro-acetylation). For peptides, there has been developed certain new, very efficient techniques, such as: electrospray ionization (ESI) and matrix-assisted laser desorption ionization (MALDI).

In the next Figure 4, is presented the mass spectrum of pure valine, recorded on a Bruker Daltonics High Capacity Ion Trap Ultra (HCT Ultra, PTM discovery) instrument.

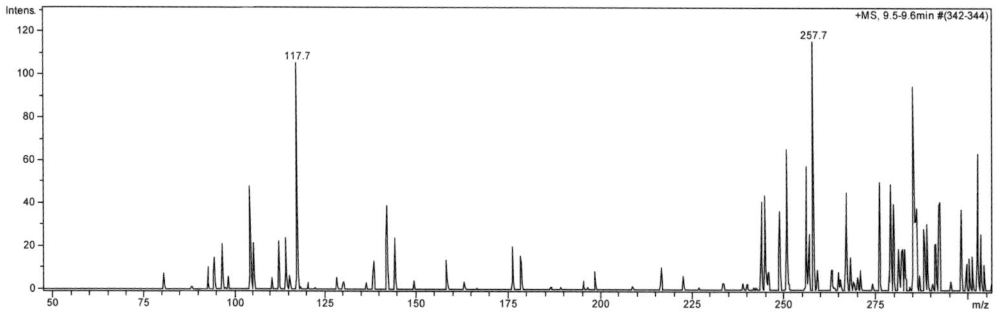

Figure 4. Valine MS-specta

NMR spectroscopy offers the most useful and valuable information about the structure of perhaps any natural product. The method has the advantage of excellent reproducibility. Even though it is considered to be one of the more expensive techniques, NMR is relatively cheap, fast sensitive and easily used as a routine application for amino acid analysis.

Author details

Ion Neda, Paulina Vlazan, Raluca Oana Pop,
Paula Sfarloaga, Ioan Grozescu and Adina-Elena Segneanu
National Institute for Research and Development in Electrochemistry and Condensed Matter, Timisoara, Romania

Acknowledgement

This study was supported by National Grant - IDEI-PCE Exploratory Research Program - Project no. **341-/01.10.2011 - Immunomodulante Fluoroglycopeptide Molecular Architectures**.

6. References

[1] M. Busquet, S. Calsamiglia, A. Ferret, C. Kamel, „ Screening for effects of plant extracts and active compounds of plants on dairy cattle rumen microbial fermentation in a

continuous culture system", Animal Feed Science and Technology 2005, 123-124, 597-613.

[2] L. Chen, Q. Chen, Z. Zhang, X. Wan, A novel colorimetric determination of free amino acids in tea infusions with 2,4-dinitrofluorobenzene, J. Food Composition Anal. 2009, 22, 137-141.

[3] S. W. Lee, J. M. Lin, S. H. Bhoo, Y. S. Paik, T. R. Hahn, Colorimetric determination of amino acids using genipin from *Gardenia jasminoides*, Anal. Chim. Acta 2003, 480, 267-274.

[4] H. Liu, S. Sun, G. Lv, K. K. C. Chan, Study on *Angelica* and its different extracts by Fourier transform infrared spectroscopy and two-dimensional correlation IR spectroscopy, Spectrochimica Acta Part A 2006, 64, 321-326.

[5] N. C. Polfer et al., „Infrared fingerprint spectroscopy and theoretical studies of potassium ion tagged amino acids and peptides in gas phase", J. Am. Chem. Soc. 2005, 127, 8571-8579.

[6] J. Oomens, J. D. Steill, B. Redlich, „Gas-phase IR spectroscopy of deprotonated amino acids", J. Am. Chem. Soc. 2009, 131, 4310-4319.

[7] T. Kasai, K. Furukawa, S. Sakamura, „Infrared and mass spectra of α- amino acid amides", J. Fac. Agr. Hokkaido Univ. 1979, 59(3), 279-283.

[8] R. Linder, M. Nispel, T. Haber, K. Kleinermanns, „Gas-phase FT-IR spectra of natural amino acids", Chem. Phys. Lett. 2005, 409, 260-264.

[9] S. Moore, W. H. Stein, „A modified nynhidrin reagent for the photometric determination of amino acids and related compounds", J. Biol. Chem. 1954, 211(2), 907-913.

[10] D. Melucci, M. Xie, P. Reschiglian, G. Torsi, "FMOC-Cl as derivatizing agent for the analysis of amino acids and dipeptides by the absolute analysis method", Chromatographia 1999, 49, 317-320.

[11] S. Kochhar, B. Mouratou, P. Christen, "Amino acid analysis bu precolumn derivatization with 1-fluoro-2,4-dinitrophenyl-5-L-alanine amide (Marfey's reagent", in The Protein Protocols Handbook, 2nd Edition, edited by J. M. Walker, 2002, Humana Press Inc., Totowa, New Jersey.

[12] E. Ngo Bum, K. Lingenhoehl, A. Rakotonirina, H.-R. Olpe, M. Schmutz, S. Rakotonirina, „Ions and amino acid analysis of Cyperus articulatus L. (Cyperaceae) extracts and the effects of the latter on oocytes expressing some receptors", J. Ethnopharmacology 2004, 95, 303-309.

[13] M. Culea, O. Cozar, D. Ristoiu, „Amino acids quantitation in biological media", Studia Universitatis Babes-Bolyai, Physica, L, 4b, 2005, 11-15.

[14] C. R. Warren, „Rapid and sensitive quantification of amino acids in soil extracts bz capillary electrophoresis with laser-induced fluorescence", Soil Biology & Biochemistry 2008, 40, 916-923.

[15] Z. Sun, J. You, C. Song, "LC-Fluorescence detection analysis of amino acids from Stellera chamaejasme L. using 2-[2-(dibenzocarbazole)-ethoxy] ethyl chloroformate as labeling reagent, Chromatographia 2010, 72, 641-649.

[16] W. Li, M. Hou, Y. Cao, H. Song, T. Shi, X. Gao, D. Waning, "Determination of 20 free amino acids in asparagus tin by high-performance liquid chromatographic method after pre-column derivatization", Food Anal. Methods 2012, 5, 62-68.

[17] M. Zhang, Y. Ma, L. Dai, D. Zhang, J. Li, W. Yuan, Y. Li, H. Zhou, „A high-performance liquid chromatographic method for simultaneous determination of 21 free amino acids in tea", Food Anal. Methods 2012, DOI 10.1007/s12161-012-9408-4.

[18] Barkholt, V. & Jensen, A.L.: Amino acid analysis: determination of cysteine plus halfcystine in proteins after hydrochloric acid hydrolysis with a disulfide compound as additive. Analytical Biochem 177, 318-322 (1989);

[19] Barth A., The infrared absorption of amino acid side chains, Progress in Biophysics & Molecular Biology 74 (2000) 141–173;

[20] Holzgrabe U., Diehl Bernd W.K., Wawer I., NMR spectroscopy in pharmacy, Journal of Pharmaceutical and Biomedical Analysis 17 (1998) 557–616;

[21] F. G. Kitson, B. S. Larsen, C. N. McEwen, Gas Chromatography and Mass Spectrometry, A Practical Guide; Academic Press: San Diego, 1996; Chapter 9.